BUSH TELEGRAPH

BUSH TELEGRAPH

Discovering the Pacific Province

Stephen Hume

HARBOUR PUBLISHING

Published by
HARBOUR PUBLISHING
P.O. Box 219
Madeira Park, BC Canada
V0N 2H0

Cover painting "Reflections," oil on canvas by Brian Scott
Cover design, page design and composition by Roger Handling
Edited by Irene Niechoda

Printed and bound in Canada

Harbour Publishing acknowledges the financial support of the Government of Canada through the Book Publishing Industry Development Program (BPIDP) and the Canada Council for the Arts, and the Province of British Columbia through the British Columbia Arts Council, for its publishing activities.

Canadian Cataloguing in Publication Data

Hume, Stephen, 1947-
Bush telegraph

Includes index.
ISBN 1-55017-215-8

1. British Columbia. I. Title.
FC3818.H86 1999 971.1 C99-910896-4
F1087.H86 1999

These stories are for Heledd
kloshe tenas klootchman, mesika tenas towah
that she might never forget the direction home.

CONTENTS

ACKNOWLEDGEMENTS

Writing is a lonely task but no book comes to fruition without much help. First, for her wise advice and tireless encouragement, I owe many thanks to Susan Mayse, whose roots run as deep as any in her beloved British Columbia. Thanks are due also to my editors at *The Vancouver Sun*, where many of these essays were first published and with whose permission they are refashioned here. They had the confidence to let me write about what I thought was important across this fabulous province. Thanks also to my editors at Harbour, Mary Schendlinger and Irene Niechoda, who whipped an unruly manuscript into shape. Finally, every writer owes a debt to those who have gone before. So here's a tip of the Tin Hat to Torchy Anderson, Arthur Mayse, Howard O'Hagan, Margaret Ormsby, Barry Broadfoot, Paul St. Pierre, James K. Nesbitt, B.A. McKelvie, Bruce Hutchison, R.M. Patterson, Terry Glavin and Roderick Haig-Brown, to whose stories one may always turn whenever one wishes to refresh a sense of wonder at the marvels of life here in Saghallie illahie.

THE PACIFICATION OF STEWART

Before most of us were born, during the high tide of north country mining camps that were about as tough as they come, they had an annual contest in Stewart. With one hand, you had to lift the pin from a milling machine, extend your arm, then hold it motionless as long as you could. The pin weighed 40 kilograms. Every year for eleven years just after World War I, Constable Lawrence Albert Newton Potterton won that contest over the thousands of hardrock miners washed up in the flood and ebb of the Cariboo, Klondike and Stikine gold rushes. A veteran of four police forces, Constable Potterton patrolled the alleys and saloons, blind pigs and fancy houses, traplines and placer camps of the North Coast and its interior from 1918 to 1952. When I met him he'd been retired almost half a century and was nearing his ninety-seventh birthday. Even with his rugged physique ravaged by time, the toughest cop in Stewart's history remained an imposing, intimidating presence.

I tracked him down at Acropolis Manor in Prince Rupert, a senior citizens' home that says a lot about community values in this fishing port. It's not tucked away in some dreary suburban cul-de-sac where inconvenient old-timers can be shuffled off and forgotten. Instead, this seniors' home reflects a community that values and respects its elders. Acropolis Manor resembles its namesake, proudly occupying the best symbolic piece of real estate Prince Rupert has to offer: a rocky crag in the centre of town with spectacular prospects of the northern port and its harbour.

When I slipped in for a visit, Lance Potterton was deeply absorbed in the controlled mayhem of a National Football League game, chuckling and offering salty asides. He listened to the TV with earphones because he was hard of hearing and spoke with a kind of thinly modulated bellow. Seeing him relish the elegant violence of his football, I could clearly tell he wasn't one of those who planned on going gentle into that good night. Next to the TV set rested a copy of Martin Cruz Smith's consummate cool cop thriller, *Gorky Park*, and a copy of Potterton's own book, *Northwest Assignment*. Lance rose to greet me from the easy chair beneath his pride and joy, the RCMP insignia and crest presented to him by the local grade seven class. He saw me admiring it and rattled off his service number, 16295. The small apartment was spartan, neat to the point of regimentation, awaiting inspection from a starchy boot camp sergeant who never came.

Lance was a rawboned six-footer. His nickname comes not from his initials, but from the impression his wife Phoebe made on the unpolished miners as she pronounced his name, "Lawrnce," with her tony British accent. "Phoebe was five foot nothing, but when she put her hands on her hips, you knew you'd had it!" he said. He reached out to shake, his hands huge and his grip still powerful, a reminder of the admiring anecdote that still does the rounds in dockside taverns. I asked whether it was true, as they told me on the docks, that he used to win bar bets by picking up nine pool balls in each hand. He grinned at the tribute and elaborated. When the Harlem Globe Trotters came through on one of their tours and heard the story of the cop with big hands, they scoffed.

They were promptly escorted to the Elks Hall for a showdown. "Not one of them could do it. I took eighteen pool balls off the table with my two hands and not one of those professional basketball guys could do it!"

Born the son of a cabinetmaker, Lance's first view of the world was a stone house in windy Yorkshire in 1894. He came to Canada in 1912, one more teenager in the British attempt to populate the far dominions of empire with a flood of emigrants. Like so many before him, he wound up punching cows on a Saskatchewan ranch. Then he weighed the long-term prospects for getting ahead in farm labour and returned to Britain. This wasn't "The Cure" taken by so many British emigrants who return home only to rediscover with speedy dismay precisely why they left the class-ridden place the first time. He already knew what he wanted to do in his new country of endless frontiers and wild country patrolled by the red coats and Stetsons of the North West Mounted Police. And he missed those clear prairie winters where you can hear an axe for miles. A slurry of ice, it seemed, had already fixed in his blood. It whispered that he was a Canadian.

That elementary voice from the wilderness still spoke to him, he said. He'd lie awake in the dark in his old man's body, listening to the murmur of an institutional building going about its sleepless business, the squeak of soft-soled nurses shoes patrolling the corridors, and then he'd suddenly be back, a young man out there on the edge of nothing—or everything, depending on the way you think of these things. "One night I went back. It was strange. I looked up and I was sleeping under the stars on the shores of Daisy Lake in the Cassiar. The lake was black as ink except for the glitter of those stars. I lay right in this bed and I could hear it as clear as anything, the rumble as huge blocks of ice fell into the lake from a glacier, shaking those stars and breaking up the stillness before dawn."

Another night, he said, he was suddenly transported back into "the black, velvet belly of the storm" that caught him on the summit of a peak above Observatory Inlet that was three kilometres high, enveloped in a huge cloud that was being bombed by lightning and thunder. The brilliant flashes came seconds apart, he said. "That mountain top was a blaze of electrified wonder."

In 1918 when he went back to Britain it was just long enough for six months at the Manchester Police Academy and eighteen months as a Yorkshire bobby—credentials he knew would win him quick entry to a Canadian police force. Then, in 1920, he headed back to God's country. "My father walked me down to the dock at Liverpool to put me on the boat. He kissed me. Then he shook my hand like a man. The last thing he said to me on this earth was the best piece of advice I've ever had from any man. He said to me: "Son, never worry about something that you can't do anything about."

Eight years earlier, as a Yorkshire lad with his head full of dreams, his plans to make a fortune as a Saskatchewan rancher foundered in 1913. It was as though the grim attrition of World War I had already begun at some deep unconscious level: immigration collapsed and a deep recession gripped the prairies. When war broke out in August of 1914, it fell with particular force upon the West, most deeply felt on the prairie cattle ranches. The biggest spreads survived, but the lifestyle on the smaller holdings suffered a fatal blow. Most ranchers were fierce British patriots of military age. When they enlisted, the kids and old men left behind to tend the cattle and horses found themselves struggling with desperate labour shortages and inflated prices. After six desperate years of hardscrabble, Lance decided if he wanted to raise a family on the Canadian frontier, he would need another career, which is why he went home for training as a policeman before courting Phoebe and persuading her to return to The Last Best West.

Their honeymoon consisted only of the leisurely train trip across the country and the not-so-private sleeping car accommodations. But at Prince Rupert, the train could go no farther and the honeymoon was over. In truth, he said, it had ended some time before.

As Lance had predicted, with his Manchester training and Yorkshire experience, he quickly found a job as city police constable. Six years later, when the city force was taken over by the BC Provincial Police, he was posted to Stewart on the Alaska border at the head of the Portland Canal. It was 850 kilometres north of Vancouver and considered the roughest mining town in BC's already rough interior. His salary was to be $2.82 per day and he was expected to patrol the back country as well as keep order in town, a duty he relished. "I loved to get out there in the bush," he laughed. "No paper work out there. No brass. The department had to issue me ten new pairs of snowshoes in my first ten years on patrol. Anyway, I had to do quite a bit of hunting and fishing to supplement my wages."

Once, shooting ducks in the estuary of the Bear River, he wounded one and it fell on the river ice. He went to get it but hadn't calculated on the falling tide. There was a half-metre gap between the water and the ice. It broke, he fell through and found that his clothes were freezing into the ice. Only his physical strength, he said, enabled him to rip himself loose. By the time he got home, his clothes had frozen solid around him.

"When I walked into the kitchen I was really inside a solid block of ice and I had to lie on the floor until my clothing thawed sufficiently. Phoebe cut off my clothes with a tin snip. Then she ran a tub of tepid water and got me into it until my body temperature came up. I fell asleep and I didn't dream at all but the next morning I woke up fresh and went back to get my gun. I found the gun right where I left it but that darn duck had gone out with the tide."

The Stewart where Lance was the only law had been born in a scam and survived in a drunken flush. It was supposed to become the transcontinental railway terminal of the North, a port for Sir Donald Mann's Canadian Northeastern Railway, which collapsed in scandal in 1912 after less than 30 kilometres of steel had been laid. Yet if commerce and the shyster politicians failed Stewart, mother nature didn't. In the post-war depression, the town boomed around gold mines like the Big Missouri, Porter-Idaho, Georgia River and the Premier Gold Mine. Premier alone hired eight hundred men. Three shifts were blasting and mucking 1,000 tonnes of rock a day. The mine paid out $4,857,000 in four years of operation, producing $16 million in total. Around Stewart, the bedrock was honeycombed with 30 kilometres of shafts and drifts.

The townsite was neatly laid out along the waterfront beneath Mount Rainey, which soars more than 2,000 metres from tidewater. During the boom of 1910 to 1913, lots in Stewart sold for $10,000, about sixteen years' pay for a constable like Lance. Weathered, unpainted clapboard buildings were scattered across the six square blocks of the townsite. If the town was a model of planning, it was not a model of good behaviour.

"I went into Anderson's Café for coffee one morning and sitting next to me is a tough-looking guy with a bushy beard and long black hair. He's wearing a carpenter's apron with a pouch full of nails and a hammer and he orders a stack of flapjacks and coffee.

"We both got our coffee but his hotcakes seemed to take a long time. He waited very patiently I thought and never said a thing. When his pancakes finally arrived they were as cold as if they'd come out of the ice box and this guy lifts them carefully off his plate, arranges them on Anderson's counter, then takes three two-and-a-half-inch spikes out of his apron and he nails those pancakes to the counter. Then he tossed down 10 cents and walked out.

"The cook came out to see what all the pounding was about. His eyes just bulged. 'That son of a bitch crucified my pancakes!' he yelled. I've always remembered that. Crucified pancakes. What a town."

There were three licensed hotels on the main street and thirsty miners jammed into them, all clamouring for beer at the same time. They ordered set-ups of three at a time in the Empress, packed in elbow-to-elbow down a mahogany bar with mirrors and a brass rail long enough to handle fifty drinkers at a set. The drinkers included men like Sir Lawrence Albert Casey, whose family had one of the largest estates in Ireland but who had been sent out to the colonies as a young man when he was expelled from university for drinking and fighting.

"He was well educated, and well bred when he was sober. He received a large remittance from Ireland every month and promptly spent most of it on whiskey and rum. Soon as he got his money, he got drunk, paralyzingly drunk, but his head was always clear. He was eighty-three then, so I'd just take him back to his cabin.

"The bad ones were the Finlanders. They would fight on the docks with their pukka knives, those knives with the horse head on the handle. They'd hold their knife right down at the end of the blade and try to cut the buttons off each other's shirts with the point. It made a godawful bloody mess but I discovered it could usually be fixed up with a few stitches."

More typical were the drifters like Peck McSwain, the itinerant printer at the *Stewart News* who had followed his vocation from boom camp to boom camp. "One day I asked him where he lived before he came to Stewart. He said, 'Nowhere—all my life I've been going from nowhere to nowhere and I never got there yet.'

"There were 4,000 men there and in 1927 and 1928 the liquor outlet in Stewart had the highest volume of sales in all of BC, probably all of Canada, maybe the continent," Lance said. "They warned me that they had four constables up there in three months and that every one of them either had to be withdrawn or wound up confined to his office, afraid to go out."

Constable Potterton grinned at the memory of bringing his wife to a place where the whole town seemed to be involved in gambling and drinking—everyone but him making big money and spending it as fast as they made it on all the seedy stuff that boomers spend on. Crap games would break out on the board sidewalks; blackjack and poker for the really big stakes went on in the "fancy houses" strung out along the edge of the tide flats.

"These fancy guy pimps with their painted fingernails—they'd play cards all night, then lay about the cathouses all day, then be out in the evening drumming up business for the women," Lance said. "I decided I didn't like these fellows very much. The only thing I could charge them with was being found in a common bawdy house. One thing I could do was remind them who the law was. But if I went and knocked on the door, some woman would yell and out the back door into the bush these fellows would go. So one day I went out back and strung a wire in the brush about ankle high. Then I went around and knocked. As soon as the girl saw me I headed for the back. This fancy guy hit the wire and then I arrived like one of those football linebackers. That was very satisfying."

Most of the women working the Stewart brothels were Americans, he said, about 10 percent of them black. Most of the black women were outstanding professional entertainers who played the piano, sang and danced. In many cases the women were mothers supporting children they'd left with friends or relatives and a surprising number were highly educated. The local bank manager told the constable that his best customers in savings accounts were the women from the red-light district, and while liquor outlets were the biggest business in town, the whorehouses ranked a close second.

"Every house had what they called a housekeeper. In most cases she was a big, middle-aged woman who could handle the girls and toss out any of the men who came to visit and got out of line. They had quite a sense of humour. I remember one of those women appearing in the back door to watch this little guy splitting wood next door. He was having a hell of a time. She stood with her hands on her hips watching for awhile, then she said, 'Shorty, I have an idea you split wood the same way I do business—upside down, crosswise, arse backwards, every bloody way but the right way.'"

If the town was tough, Lance had learned a few unofficial lessons from the cowboys in those prairie bunkhouses. He and his elegant young wife descended on Stewart like a whirlwind of bush justice. To those who wanted to take him on—and there were plenty—he'd clench those fists that could hold nine pool balls and reward them with a fearful beating. Then let them sober up in the jail and turn them out in the morning—sore, hungover and remorseful. "Oh yeah, it was a pretty tough town. Or it was for a couple of months. At first I had quite a bit of fighting to do. I won all of them. That was what earned the people's respect. After that it was pretty quiet."

And Phoebe's contribution to the pacification of Stewart? Culture for the uncultured. If the police building was primitive—office in front, jail behind, a kitchen, living room and two bedrooms between; wood heaters, no bathroom, no hot water—it served her other purposes. After Lance dealt with that first crush of business, Phoebe took advantage of the eleven-year lull that followed. She used the unused jail to teach the local girls ballet.

Lance's son-in-law Foster Husoy, long retired from the seiner fleet and old enough to be my own father, tells me another story about how peace and tranquility came to the streets of Stewart as we head up the hill to Acropolis Manor. He's not sure Lance will want to talk about it.

"Well, there is this guy who prides himself on being the meanest, toughest in Stewart. He's called Big Jock. This is about 1921. Jock is getting ugly in the tavern one afternoon when Lance comes in. He tells Jock it's time to come outside like a man and settle this where nothing will get broken. Out they go to the main street of Stewart with everybody watching. This guy hits Lance with his best lick and staggers him.

"But then it's Lance's turn. He was hard as iron when he was a young man and he knew where to hit. He punches Jock right under the heart and knocks him flat. After that, no one in Stewart ever really challenged Lance again. Every time somebody got the idea of breaking things up in the bar, Big Jock would step in. 'Only after you take me,' Big Jock would say. 'The cop took me. If you can get through me, then you get your crack at the cop.' Not many tried to get through Big Jock and nobody ever did."

In the decade after Big Jock, the provincial constable had only one more fight in Stewart. That was with a guy who had welded brass knuckles to a knife blade. Lance got cut, but he took away the weapon and put the drunk in the cooler to wait for the next boat out.

Lance snorted when I raised this bit of family folklore. He thought it inconsequential and preferred another story. This one is the mystery of the Stewart prospector who in 1927 stumbles into an old drift driven into a promising copper outcrop. There are no signs a claim had ever been staked, so the prospector goes in to look for some assay samples. Deep in the gloom of the drift, he finds two bodies. "They were mummified," Lance said. "Mummified! Those guys were the original prospectors. It looks as though they had killed one another in a fight over something we'll never know. Copper salts leaching out of the deposit pickled the bodies where they lay."

When I expressed disbelief at this bizarre tale of BC mummies, he slapped the back of his memoir, *Northwest Assignment*. It's a book loaded with the meticulous detail you'd expect from a man who spent half a century preparing reports for four different police forces. "I only put in my book what I thought people would believe," he says. Then he extracted some photos from a yellowing file. It was grisly proof of those modern day mummies, their preserved skin dyed the blue-green that copper goes when exposed to the weather. "Couldn't put this in the book now, could I?" Subsequent investigations narrowed the mystery down to two prospectors who went missing in 1897 while the world's attention was turned not to copper, but to nuggets in the Klondike, he said. "Still, I've seen worse."

And what might be worse than that?

"Watching Eneas George and Richardson George of Lytton hang on March 8, 1936 for the murder of two provincial policemen. They hung those boys next to a bank robber named Charles Russell. I thought it would have been an act of grace to stay their execution. We'd given the Indians tuberculosis and other diseases, whiskey and gin, and we still had something left to give—the rope.

"Let me tell you, son, it's a hell of a way to kill a man. I've seen it. I watched them drop—those poor boys jumping and kicking for ten minutes afterward—just nerves they said. Man."

I asked if he'd ever encountered Simon Gunanoot, the legendary Gitksan outlaw from Kispiox who eluded the West's biggest manhunt for thirteen years between 1906 and 1919 following the murders of Alex McIntosh and Max Leclair. Gunanoot's father warned him he'd never find justice in a white man's court so he took to the bush. He was said to be able to cover fifty miles a day on foot in the heaviest underbrush.

"Did I know him? I had him in jail a couple of times," Lance said. "He was a tall man, over six feet, stood straight as an arrow. He was strong and he ran like a deer. Simon had a personality all of his own. There never was another Simon Gunanoot."

In 1906, while Gunanoot was still a fugitive, his father Nahgun died. At the height of the manhunt, Simon slipped silently through the posses and police lines to fulfill his father's request that he be buried on Bowser Lake. Simon packed his father the 65 kilometres to Graveyard Point and buried him there as he'd been asked.

"Well, after he surrendered and was acquitted on October 8, 1919, Simon and his sons Charlie and David ran three 150-mile traplines from Portland Canal into the headwaters of the Nass and Skeena rivers. They brought in some fine furs and a fur buyer from

Vancouver went up. The boys had been drinking. When this fur buyer tried to skin them on the deal, they beat him up and threw him out of camp.

"I was called to settle things down. Simon was some distance from his tent. When he saw me, he edged towards the tent. I noticed his rifle leaning on the tent, so I waited. He kept moving. I drew my revolver and shot the whiskey bottle near his feet. I said: 'You're next, Simon.' When he saw that glass fly, he smiled, but he backed off and we cooled everything down."

I looked skeptically on this tale. Lance laughed again, rummaged in a drawer and extracted a faded target. The official tag said: "Perfect shot, June 26, 1940 with .38 calibre service revolver at 25 yards." I examined the bull's eye. Ten bullet holes clustered within a centimetre of dead centre. Lance Potterton, it turns out, once ranked third with service side arms among all the marksmen of North America.

"I told you," he grinned, "I only put in my book what people would believe."

Desert

Just as daybreak splashes the mountains with rose and amber and dawn begins to glimmer in the lake below, the eye first deciphers the delicate calligraphy of rattlesnake tracks—but only just. In the silvery dust and sagebrush 400 kilometres east of Vancouver and just outside the town of Osoyoos, the faint, articulated patterns left by pit vipers seem vaguely high tech, resembling nothing so much as the treadmarks made by some kind of indescribably light mountain bike. The feathery ridges kicked up by their moving ribs and scales are best discerned in the relief cast by the brief, slanting shadows of early morning.

If daybreak is the best time to examine rattlesnake tracks, it also seems like the only time. At solar noon the temperature on these exposed flats routinely soars to a brain-frying 48 degrees. For this is Canada's true desert: a dessicated finger of the great Sonoran desert that reaches up through the rainshadow all the way from Mexico; a stark, mummified reminder that the real geography of this continent runs north-south and mocks the temporary east-west scratchings of political cartography. Technically, this tawny expanse of grasses and drought-blackened scrub is an antelope-brush ecosystem, although here in the south Okanagan they like to call it their "vest-pocket desert." Lionel Dallas snorts at that description.

"The pocket desert—that's a concept we have to get away from," he says. "This isn't isolated. It's part of a tongue of desert that connects us with the southwestern US and Mexico. It's all connected."

Lionel is active with the Osoyoos Desert Society and to him this subtle distinction is far more important than the tourist come-ons and cute phrases of promotional brochures. The antelope-brush ecosystem is also one of Canada's most fragile and endangered landscapes, a dwindling residual thumbnail of something that was never extensive to begin with. The parched grasslands in which this desert ecology occurs comprise only 0.3 percent of British Columbia, virtually all of that restricted to a few interior valleys in the eastern rainshadow of the Cascades and the Coast Range. Of that, only 0.2 percent—around three one-hundredths of one percent of the province—is this true desert. Most of that has already been destroyed or severely damaged by the relentless demands of irrigation agriculture, industry and urban sprawl.

Lynne Atwood, a Desert Society habitat biologist who's working on a project that hopes to restore some representative samples of the ravaged ecosystem, says that 60 percent of the original desert habitat has disappeared; 30 percent of what remains is on First Nations and private land and beyond the reach of conservationists. Only 3 percent is in pristine condition.

Ruth Schiller, sixty years an orchardist in the south Okanagan, acknowledges she is among those to blame for the vanishing desert. "When the orchards were planted nobody cared about the desert," she says. "If you saw cactus, you just tore it out. We've messed it up in every way possible. But the older you get, the more you understand." Today she's president of the Desert Society and fiercely committed to the battle to preserve for future generations what little of the desert remains. "It's disgraceful that it's threatened like this," she says. "This is a perfect opportunity for the ministry of environment, which is in a lot of trouble for what it hasn't done, to look good doing the right thing, the thing that needs to be done."

If farmers and developers have tended to perceive it only as an arid waste inhabited by venomous snakes, scorpions and black widow spiders, Melinda Misener, a visitor from Kelowna who hiked in to experience the place first-hand, says her fascination is with the "absolute contrast" between the luxuriant, highly modified industrial landscape of resorts, orchards and vineyards and the empty austerity of the desert.

Biologists, on the other hand, say the emptiness is an illusion. The desert is actually a teeming storehouse of unique and irreplaceable life and these tattered remnants hold an enormously significant biological inventory, Atwood says. To walk the land with her is to experience a kind of revelation as she distinguishes between the different types of sagebrush—"that's a Threetip sage, that's Great Basin sage"—and the astonishing array of grasses and plants. There are Red Three-awn and Dusky Maiden and greasewood, an amazing shrub whose seeds can lie dormant in the soil for two hundred years until they are triggered into germination by a passing wildfire. This is a world of dwarves and miniatures, of highly evolved economies and efficiencies where everything's adapted to the conservation of water and its sudden exploitation. Some plants are almost invisible. Atwood kneels to examine what appears to the untutored eye as a thin rime of discoloration on the soil. "Rusty steppe moss. It dries right down to this, but then, with the slightest exposure to moisture it just explodes. It plumps up and goes vivid green and a furious level of photosynthesis starts almost instantly," she says. "Winter is the season of the mosses. They absorb the water and splash the desert with green."

If she is passionate, perhaps it's because of her graduate

research at UBC, where she studied the moisture retention in soils beneath mosses. "People don't see a lot of the beauty in the desert, but look how this moss is up on tiny pedestals," she says. "It's holding the soil and preventing erosion. If this area was covered with this nearly-invisible moss you would have much less soil erosion."

More than 100 rare plants and more than 300 rare invertebrates—from the stingless sun scorpion to the Mormon Metalmark butterfly—are confined to this habitat. One in five of BC's endangered or threatened vertebrate species is found here. Sixteen of the species appearing on BC's "red list" of threatened or endangered species rely on this ecosystem. They include the tiger salamander, the burrowing owl, the northern long-eared pallid bat, the sage thrasher and the upland sandpiper. Another forty species native to this landscape are on the province's "blue list" of those considered vulnerable and at risk. They include the painted turtle, the badger, screech owls and turkey vultures, the burrowing spade-footed toad, the world's smallest hummingbird—and, of course, the western rattlesnake.

Consider this particular rattlesnake highway, for example. It's a route that the snakes have followed for thousands of years, travelling down in search of water from the scorching benchland where they hunt, bask or seek shade to regulate their temperature and make the winter dens that permit them to survive the cold winters. Their habitat is a kind of time machine, a snapshot from 8,000 years ago when a great warming seized the earth and the desert and its creatures extended their reach up the hot valley bottoms and into much of BC's southern interior. Those ecosystems began to contract again during a cooling cycle 5,000 years ago.

Now the rattlesnakes' timeless journey to drink at the lake brings them into conflict with an encroaching human population. Run by the Osoyoos Indian band, the lush, manicured Nk'Mip campground sprawls along the lakefront on the eastern fringe of this dry, sun-drenched border town of 4,100 which basks both in the heat shimmers and in its pride at being chosen Canada's most beautiful small town in 1995. With the warmest fresh water lake in Canada at its doorstep, cloudless skies and balmy, bug-free evenings, the spectacular 225-site campground is a justifiably popular resort destination. It's also adjacent to one of the few relatively pristine sections of desert left, most of it on the Osoyoos band's reserve. And it straddles the rattlesnakes' ancient route to the water.

Just past the gates, a sign indicates the confrontation in which rattlesnakes and campers are unwitting players. There are safety tips for dealing with snake encounters and the reassuring news that one of Canada's leading snakebite specialists resides in Penticton, about forty minutes of fast driving up the valley.

This has been a good year for rattlers acknowledges Lana Hall, who works at the campground's front desk. There have been more rattlers than she's seen in years, she says—up to three or four a day on the campground. She is cheerfully pragmatic about the snakes and the daily patrol that takes them to safety. Indeed, the concern is as much for the snakes as for the campers. "You can call it our official catch-and-release program," she says, a sly allusion to the wet side of the mountains, the west coast fishery and its on-going problems. And while the snakes demand caution, the increase in sightings is actually good news. While the western rattlesnake is British Columbia's largest—and only venomous—snake, it's also a timid, retiring creature that's been made the victim both of bad press and of changes to its environment. Highways take a heavy toll on the slow-moving snakes. And the ignorant still assume any rattler they encounter is a threat that should be killed—although the risk of being bitten is considerably less than the risk of being struck by lightning. Nonetheless, western rattlers have been hunted to near-extinction across large parts of their historic range.

It's to counter misunderstandings like these that the Desert

Society embarked upon an ambitious $5.5 million project to establish a desert interpretive centre on 83 hectares beside Highway 3 and across the valley from the Osoyoos reserve. When it is completed plans call for educational natural history displays, collection of wild seeds for future restoration work and, in the big dream, greenway corridors that link surviving nodes of desert. Eventually, the society plans to rehabilitate the immediate desert area which has been damaged by road construction, the encroachment of irrigated orchards, heavy grazing by cattle and an accompanying invasion of Dalmation toad flax, cheat grass, mullein and other aggressive non-native plants that choke out indigenous species. The restoration phase is Lynne Atwood's responsibility—which is why every baking summer day, even at 48 degrees, she is out at first light supervising a crew that braves the snakes, ticks and scorpions to weed the fragile desert by hand.

Snow

In January snowdrops shine like tiny pearls of light in West Coast gardens and even luminous hyacinths bloom here and there. Already the assumption of imminent spring resonates in conversation throughout the temperate bubble inhabited by Lower Mainlanders. And yet, a few scant paces beyond the balmy habitation sphere that envelops the Fraser River flood plain, the creeks are clad in ice, a record snowpack deepens with every precipitation-laden storm cell passing inland from the Pacific and the mountains groan and rumble with avalanches. The snow falls on Greater Vancouver, too, but for all its temporary inconvenience it passes quickly as a shadow through our days and we consider it a seasonal oddity. Lower Mainlanders go up into that other country to work or play but somehow it's always imagined as distant, exotic, even dangerous. Some of us leave on an afternoon hike outside the bubble and never return, swallowed up in the measureless landscape of winter.

"Mon pays, c'est l'hiver," sang Quebec's popular poet laureate Gilles Vigneault in a folk anthem that speaks to so many because it's true for so much of Canada: "my country, it is the winter."

Indeed, beyond our tiny patch of warmth, what Voltaire dismissed as "a few acres of snow" fans out from Hope to Kelowna and onward into a frozen province the size of western Europe. From there the snowfields extend into the Arctic territories and across the pole itself, 9.9 million square kilometres of snow, add another 14 million square kilometres upon the sea ice of the Arctic Ocean, east to Greenland's 2.1 million square kilometre ice cap, west to Alaska's 1.4 million square kilometres, the 13.4 million square kilometres of Siberia and then on across Russia to the tundra of Finland and Norway. These numbers are almost too big for the mind to encompass. German poet Walter Bauer found in this immense, unimaginably vast expanse of snow "the sum total of all wisdom. Silence. Nothing but silence. The end of time." Or, at least, the beginning of time as we know it. There is snow at the bottom of Canadian glaciers that is more than 50,000 years old, snow that fell during the last great ice age, perhaps upon the first human beings on this continent. In fact, in his book *Blame It on the Weather*, Environment Canada's senior climatologist David Phillips calculates the total number of snowflakes that have fallen over the history of the planet:
100,000,000,000,000,000,000,000,000,000,000,000.

"It snowed and snowed, the whole world over, snow swept the world from end to end," writes another poet, one of Boris Pasternak's winter-bound characters in *Dr. Zhivago*. "A candle burned upon a table, upon a table, one candle burned." Perhaps it's this sense of huddling in our own small Eden in the midst of entropy, some sense of that Russian poet's warm globe of candlelight crowded by snow and darkness, that generates the self-absorbed ignorance toward conditions in the rest of the province that so many British Columbians from beyond Hope profess to feel radiating from Greater Vancouver. If two men meet and they both understand snow, snorted poet Carl Sandburg, who as a Chicago resident was no stranger to it himself, you know they're both

Canadians. If one of them doesn't, a wag from Fort Nelson might add, you know which one's from Vancouver. And yet there are forty-three place names in BC that refer to snow, a category exceeded only by those names referring to rock. Snowdrift, Snowbank, Snowball, Snowslide, Snowpatch, Snowsquall, Snowsaddle, Snowtop, Snowwater, Snowman, Snow Dome, Snowcap. The list goes on, reflecting a powerful reality that somehow fails to impinge at Granville and Georgia.

Snow, however, is vital to the economic well-being of the province and not just as the engine of a multi-billion-dollar winter tourism and recreation industry. It's critical to forestry, to the fishery, to rural agriculture and suburban horticulture and to hundreds of communities that rely on surface runoff to replenish their reservoirs of potable water. Snow serves as a kind of precipitation bank, releasing cool water into the rivers during long hot summers in the cordillera—rivers that provide a highway for temperature-sensitive salmon bound west, and that spill eastward to irrigate the great breadbaskets of Canada and the United States. To calculate its importance, consider that one tonne of alfalfa represents an investment of over 900,000 litres of water and that virtually all the water available for agriculture and domestic use on the western half of the continent originates as snow.

But summer visitors from the Lower Mainland to a remote village like Wells, high on the crown of the Cariboo Mountains, 700 kilometres north of Vancouver, will ask ingenuously why so many doorways are set so high above the ground. Because the snow can be deep, the inhabitants wearily reply. "Deep" is an understatement. In the Coast Range, snow accumulations regularly reach depths of 11 metres or more. At Mt. Revelstoke it once piled up to a depth of 24 metres. In some places, assisted by wind, it can reach depths of 35 metres. The Lower Mainlander's mystification with these matters is easily understood. For instance, the combined annual snow removal budget for Surrey and Coquitlam is $1 million, an amount that pales by comparison to the $30 million required for Edmonton and Calgary. Add Regina and Winnipeg and the total is $50 million. And in Moscow the bill runs to $4.4 million per day. West Enders may be dithering over whether it's time to switch from splash suits to tank tops for that jog along the sea wall, but everywhere else winter still rules what writer W.E. Collin called "the white savannah," that dazzling expanse of snow that cloaks the northern hemisphere, a bond that physically unites British Columbia with Irkutsk and Thule, not with southern California as in popular Lower Mainland mythology.

Thirty-six percent of Canada's total precipitation falls as snow; world wide it's only 5 percent. More than 90 percent of the province is covered by snow between the equinoxes. On the Lower Mainland, for the most part, the connection is invisible—but it is there nonetheless. What usually arrives in Greater Vancouver as rain sifting out of a grey sky is actually piling up above the cloud line in the billions of tonnes. In one ten-day period in 1996, for example, it is estimated that 600 million tonnes of snow fell on the region. In fact, although we choose to perceive otherwise, winter often exerts its authority. With the facets of all those tiny crystals aligned to form a planetary-scale mirror, snow reflects 87 percent of the sunlight that reaches it back into space, profoundly influencing the world's climate and weather patterns in the process. And in winter the influence on Lotus Land can be immediate. Super-chilled Russian air that meteorologists call the Siberian Pipeline squirts down through Alaska and BC's interior valleys to block the warm, moisture-laden air from the south Pacific we call the Pineapple Express. When they meet, the collision is titanic. It's why Victoria is actually one of the snowiest places in Canada, the only city in western Canada to have had more than 65 centimetres of snow in a single day, sometimes receiving as much as 78 centimetres in a day-and-a-half.

The Garden City prefers another statistic: on average it has the smallest annual snowfall in Canada. Some meteorologists, however, point out that, using statistical averages, the odds are that there won't be a truck on the freeway in the precise spot at which you decide to cross—but common sense suggests it's not a good idea to step off the curb. Perhaps this refusal to perceive the real nature of the physical landscape surrounding us is not surprising. Snow is one of the great shape-shifters of the natural world, an elementary presence that represents the magical passage of water from gas to solid without passing through the liquid state that lies between.

Snow comes in many forms—particles as dry and fluffy as powdered milk that draw skiers to the Bugaboos, or flakes the size of pie plates that were recorded in Montana in 1887. Snow has different characteristics according to the region in which it falls. At more than 400 kilograms per cubic metre, snow in Greater Vancouver weighs more than twice what it weighs in Winnipeg, points out Phillips. So you're more at risk of suffering a heart attack from clearing the drive on the North Shore than are your cousins in Moose Jaw. New-fallen snow may be 97 percent air, but shovel a boot-high layer from the sidewalk in front of a typical city lot and you'll have moved six tonnes.

The Inuit, for whom snow is not an enemy, but a friend—a highway, a house, insulation, a silent language by which they interpret the natural world—have more than thirty different words for describing its characteristics. *Nutagak* is fresh snow in which you can look for tracks. *Aniu* is packed snow—you can cut blocks for an igloo. *Mitailak* is deceiving snow, covering an open spot in the ice. *Sisuuk* is snow that will bury you in an avalanche. *Auksalak* is melting snow—you need your kamiks made from seal flippers, for wet feet freeze. *Kaiuglak* is snow that is hard and wind-rippled like sand on a beach. *Kannik* is a single flake. A snowflake forms when one molecule of water vapour in super-chilled air attaches itself to a mote of dust in the atmosphere. It grows by attracting and absorbing other molecules of water vapour which are expressed as symmetrical crystals. In many ways, the true beauty of snow resides in the symmetry of its opposites: each flake is a unique crystalline structure, a tiny brittle cosmos unto itself; combined, they provide a unity of natural architecture, the sweeping panorama of wind-carved cornices and glaciers that Byron called the diadem that crowns the mountain ranges.

Snow falls without discernible sound—from the stars, say the Inuit, drifting earthward with the returning souls of the dead through holes rent in the fabric of heaven—but when the wind blows, it whispers. It has the capacity to adhere to almost anything and yet in dry cold it will flow like a river: stand on your partner's shoulders in whiteout conditions at ground level and above it the sky will be clear and black and the stars blazing. As salty urban slush or lethal backcountry avalanche, we curse it. And yet snow also brings a healing hand to the earth, rendering soft and uniform the hard, individual edges of the world, swaddling even our most tormented industrial landscapes with infinities of white.

George Orwell, searching for a metaphor to describe the kind of bureaucratic euphemism political apologists wield when they want to talk about the indefensible without evoking any mental image of the subject matter—"pacifying" the village when they really mean incinerating its women and children with napalm; "sanitary landfill" when they mean garbage dump—chose the image of snow for its uncanny ability to camouflage what is ugly. "A mass of Latin words falls upon the facts like soft snow, blurring the outlines and covering up all the details," he wrote in his chilling essay "Politics and the English Language." This ability to soften and hide the evidence of our excesses causes us to imbue the idea of snow with romantic qualities. In popular depictions we tend to associate it with festive holidays and childhood fun, not slush and heart attacks and fatal traffic accidents.

Yet, like life itself, snow proves ephemeral, soon melting into memory, especially in coastal BC, where its appearance is always a matter of chaos and delight and front page headlines that leave them shaking their heads in Atlin and Lower Post. It's winter. It's Canada. It snows. And then it goes. The spring that's already unfolding in the Fraser Valley will soon climb to the snowy uplands with a blizzard of vivid wildflowers hard on its heels. The glittering snowpack will begin to melt away into the memory of one more winter past. But as all memories do, it will continue to nourish us until what made it comes again.

SKUNK CABBAGE

As April embroiders the woods with a fragrant new filigree of green, our pathway into spring is suddenly illuminated by one of the world's small miracles. Bursting from the black muck of alder bogs and river sloughs, like yellow flames, charged with a furious sexuality, comes the tightly furled perennial that we disdainfully dismiss as skunk cabbage. *Lysichitum americanum* is the proper botanical classification for BC's skunk cabbage. *Lysichitum camtschatsensis* is its almost identical Asian cousin. *Symplocarpus foetidus*, a different genus, is the east coast's version.

Others don't share our disdain. In Europe, skunk cabbage is prized as an ornamental. In Japan, when skunk cabbage first appears in the great bogs of Hokkaido, 10,000 visitors a day come to see it and celebrate spring. But the real marvel of the aroids, the plant family to which skunk cabbage belongs, is not their beautiful blooms but the ability of some members to behave like mammals, generating and regulating their own heat regardless of external temperatures. Science has known about this phenomenon since the Age of Reason and folklore has tracked it much longer than that, but there is now a renewed interest in the physiology that permits it.

Mention skunk cabbage to non-scientists, however, and you'll generally draw blank looks or wrinkled noses, not without reason. Some species more potent than the one found in BC put out a perfume that's described by Gordon Leppig of the California Native Plant Society as "a rank, fetid, odiferous, putrid, vile, revolting, vomitus stench." Conservatories wisely display these botanical wonders near special emergency exits, he points out, so that visitors overcome by the putrescent miasma can flee to do their retching outside. Still, our disrespect for the milder BC genus seems misplaced. The western skunk cabbage once held a place of importance among West Coast aboriginal societies, where it warranted its own stories and a special dance mask for winter ceremonials. Skunk cabbage leaves are still called "Indian wax paper" across the Northwest Coast because of their usefulness in lining berry baskets and steaming pits and in packaging preserved food. The Bella Coola used them to make ingenious disposable, biodegradable drinking utensils. The Squamish made sun shades. The Gitksan made an anti-hemorrhagic medicine, the Clallam one to treat skin lesions; the Cowlitz made an infusion for the treatment of rheumatism. Even the newcomers, faced with a shortage of European doctors, adopted it as a common treatment for ringworm. Leppig points out that New Age herbalists now list it as a remedy for bronchitis, asthma, epilepsy, hysteria, hydrophobia, lockjaw, lumbago and suspended animation. Suspended animation? Well, the smell of some species might prove sufficient to raise the dead and make them flee. And yet, research by a team of seven scientists at UBC in 1995 reported in the *Journal of Ethnopharmacology* that compounds extracted from skunk cabbage displayed non-toxic anti-viral properties when applied to the Herpes 1 virus. It seems the Clallam knew something when they used it to treat cold sores.

Young spring skunk cabbage leaves make an agreeable pot herb for the bold forager for wild greens. But if you sample them, make sure that you change the water several times while boiling them. Survival guides tell downed pilots what Native Indians already know: in times of human famine, skunk cabbage roots provide excellent emergency rations. They can be roasted and ground into a hot, gingery flour, although they require prolonged cooking. University of Victoria ethnobotanist Nancy Turner warns in her catalogue of edible plants that skunk cabbage is characterized by the presence of long, sharp crystals of calcium oxalate. Unless neutralized by cooking, these crystals puncture mucous membranes in the mouth, causing severe inflammation and an intense burning sensation. Perhaps this is why the scientific name for the plants' genus—arum—is derived from the Arabic word for fire.

For most of us, skunk cabbage now has a strictly aesthetic value. From urban Vancouver ditches to rain forest seeps in Pacific Rim National Park, in spring almost every mucky place is brilliant with their vivid yellow sheaths. The colourful hood or spathe protects a thick, fleshy, decidedly phallic-looking stalk called a spadix. Technically, this is not a flower but an inflorescence—thousands of tiny blossoms clustered to resemble a larger bloom.

Yet what makes the skunk cabbage most distinctive is not the sickly sweet scent for which it is colloquially named: that odour, vaguely reminiscent of carrion, appears to serve as an attractor for early pollinating insects. Nor is it the exotic tropical look of its enormous, waxy foliage. Related to the Polynesian taro plant, the skunk cabbage has leaves that are among the largest of any native North American plant, growing up to 1.5 metres in length and spreading like huge wrinkled elephant ears. What interests scientists about the skunk cabbage is its family's ability to generate and regulate temperature. For example, just as new growth shoulders up into the frosty air of February, the temperature inside the still-furled helmet of the eastern skunk cabbage has been reported to reach 35 degrees. Some aroids have been recorded at 46 degrees.

University of Adelaide zoologist Roger Seymour, writing in *Scientific American* about the scholarly interest in the thermoregulatory capacity of these plants, says that the eastern genus of skunk cabbage known as Jack-in-the-Pulpit reportedly will melt the snow around its emerging spathe. Seymour's research found that heat production by these remarkable plants in some cases surpasses the abilities even of animals with high metabolic rates—hummingbirds in flight, for example. This is significant, Seymour writes, precisely because it shows "striking similarities between animals and plants, two groups of organisms usually considered to have little in common." Aroids generate their heat in two ways, either by oxidizing massive stores of carbohydrate in the root or by burning lipids, fatty substances more commonly associated with animal metabolism. "Why might plants thermoregulate?" Seymour wonders. "In birds and mammals, temperature regulation provides the consistent warmth that cells need to carry out biochemical reactions efficiently. Warm-blooded, thermoregulating animals can therefore be active and continue to seek food when cold weather slows the cellular reactions, and hence the activity, of such cold-blooded animals as reptiles." One reason for the plant's unusual physiology might be that by maintaining an internal temperature which ranges from 15 to 22 degrees during February and March, the plant creates a controlled climate which is somehow critical to the development of its sensitive reproductive organs, Seymour suggests.

Demi Brown, author of one of the definitive works on the subject, says the internal temperature of skunk cabbage and other members of the arum family actually vaporizes the scents which attract pollinating insects. The spathe of the western skunk cabbage even produces a chemical which intensifies these smells and makes them last longer. Once attracted inside the spathe, the warmth of its microclimate will often incite these insects to wild sex orgies of their own. "Some have been seen with nearly 200 beetles covering the spadix and filling the chamber with a orgy of mating and feasting on the nutritious sterile flowers and secretions," Brown says. "In the melee the beetles are smeared with resin that oozes from the spathe, especially at the constriction which they have to squeeze past, and so when the male florets release pollen in the final stages of flowering, it sticks easily to their hard, shiny bodies." Seymour concurs. A stable temperature warmer than the surrounding air might well make visits easier and thus more attractive to large pollinating insects like bees, which typically expend great amounts of energy staying warm while flying.

All these hypotheses are the trigger for renewed interest in a largely unappreciated native plant. The rest of us, doubtless, will be content with the seasonal spectacle of these elegant swamp lanterns blazing away in the forest gloom. After all, it is nature's radiant signal that winter has been banished for one more season and life is about to renew itself in all its sticky, fecund, dishevelled bliss.

Invasion Biology

Seldom larger than the palm of a woman's hand, its mottled carapace ranging from green to reddish hues, the European green crab is an astonishingly attractive creature. Native to European shores from Morocco to Norway, the crab grows to about 7.5 centimetres in width and is easily identified by five spines at the front of its carapace and the flat back legs which enable it to swim. *Carcinus maenus* is of particular interest because it is unusually smart for a crustacean. As a result of its abilities to travel and learn, the $200-million-a-year shellfish industry in Washington and British Columbia are now considered at serious long-term risk from an ecological invasion by this small, non-native swimming crab. This pretty little European colonist is a fierce competitor of the big, meaty, red-shelled Dungeness crabs that are sold live in up-market seafood restaurants. In addition, the green crab hunts and devours a wide range of native shellfish species, some of which can command up to $400 a kilo in the fresh carriage trade market.

Ocean scientists are tracking the green crab's invasion of the Pacific coast with a mixture of awe and alarm. "This could affect the way in which we do all our intertidal aquaculture," warns Glen

Jamieson, a crab specialist at the Pacific Biological Station in Nanaimo. "Its main prey seems to be bivalves [which include commercially valuable clams, oysters, scallops and geoducks] and there's nothing you can do to get rid of it."

The study of the large-scale movement of species and their potentially profound impact upon evolution and biogeography is a hot new field emerging in the life sciences. It is called invasion ecology and proponents of the new discipline are particularly interested in what appears to be a worldwide alteration of naturally evolved ecosystems. Some of these changes are deliberate: the introduction of higher-yield Canadian conifers to Europe, for example. Some have political overtones: a decision to introduce Atlantic salmon to the Pacific was made despite objections from the British Columbia Transplant Committee. Presumably, potential commercial value was deemed to far outweigh any potential risks to existing ecosystems. And some changes, of course, occur by unwitting error or by accident. Whatever the reasons, many modifications are now occurring on a continental scale and at a pace that seems unprecedented in the planet's evolutionary record.

Ecosystems are dynamic, of course, and have always changed as different species evolve, migrate and seize advantage in the competition for niches in the complicated hierarchies of life. Without such change, we humans wouldn't be here to think about it. Indeed, BC itself is largely occupied by plant and animal invaders that came by both land and sea, point out Verena Tunnicliffe and Melissa McQuoid, two highly respected scientists at the University of Victoria. The large mechanisms by which new species are introduced from one region to another are relatively simple. Continental drift either isolates populations or brings them together. Climatic change alters range availability for differently adapted species. Finally there are what scientists call "unusual transport opportunities"—hypodermic syringes exchanged by drug addicts that provide a vector for the viruses which cause AIDS or hepatitis, for example, or the fouled bottoms of supertankers which carry organisms around the world.

When the Fraser ice sheet advanced and retreated from BC between 30,000 and 10,000 years ago, shorelines changed radically, land corridors and new climatic regimes emerged. These factors all facilitated species interchange between Asia and the Americas and back and forth between North and South America. Today scientists have determined that while less than 10 percent of North American mammals have southern origins, over 50 percent of South American mammals have northern origins, Tunnicliffe and McQuoid wrote in a discussion paper entitled "Invasion Dynamics and Exotic Species Introductions." All these changes occurred relatively slowly, taking place over millennia.

What's at issue is the speed with which human technologies of commerce and industry appear to be transforming large and complex ecosystems which took millions of years to develop. These changes can have economic as well as ecological consequences. When zebra mussels were imported unintentionally to the Great Lakes their population suddenly exploded, colonizing and eventually choking the expensive plumbing of industrial plants. A species of jellyfish has recently invaded the Black Sea, displacing more than 80 percent of the indigenous species in the process and ruining commercial fisheries. Some of these invasions have had a high media profile. The introduction of sea lampreys into the Great Lakes via the St. Lawrence Seaway caused millions of dollars damage to valuable sport fish stocks and galvanized public attention. And hybrid Africanized honeybees that escaped from a Brazilian lab forty years ago were sensationalized as "killer bees" because they are more prone to sting aggressively. Their migration north from South America has been steady. By displacing more easily handled European bees as they advance, the Africanized bees threaten to add costs to North America's domestic honey industry. More recently, escapes of Atlantic salmon from west coast fish farms and the discovery that some appear to have spawned successfully in an important river have triggered furious debate about the safety of fin fish aquaculture in the open ocean.

Some invasions have been decidedly low profile. A mud snail introduced to California from the Atlantic coast with the importation of eastern oysters between 1901 and 1907 has successfully displaced the indigenous species. A predatory north Pacific starfish which now infests Australian estuaries has pushed a native species, the spotted handfish, onto the endangered list by devouring its eggs. In British Columbia's Lower Mainland, a non-native species called purple loosestrife, which generates a million seeds per plant and invades fish restoration areas, is now present in twenty out of fifty sites surveyed in the Fraser River estuary. To be nondescript is not to be ineffective. Purple loosestrife is responsible for the conversion of more natural wetlands in BC than all human development combined.

On the Atlantic coast, scallop ponds are choked by a bottom-smothering Japanese alga and the green crab has invaded the once-lucrative scallop fishery off Martha's Vineyard. Municipalities in the region now pay fishers, who once harvested $22 per kilo scallops, a $1.14-per-kilo bounty to clear the crabs off the scallop grounds. Behind the immediate financial implications for BC's economy that are posed by the advance of the green crab into west coast Canadian waters loom much bigger issues. The European green crab's destructive beachhead in the commercially valuable shellfish beds of the Pacific appears to be one more signal that the rate of change to large natural ecosystems is accelerating. For example, the green crab was first identified on the West Coast less than a decade ago. Already its range has extended half the distance from San Francisco Bay, where it was first reported, to the rich and potentially vulnerable waters of Puget Sound and BC's south coast. The first specimen identified in BC waters, found by a beach walker in 1999, was hiding under a clump of seaweed at the head of

Useless Inlet in Barkley Sound. It probably arrived in microscopic larval form, carried north with the currents fuelled by the 1998 El Niño, the periodic northeastward surge of a vast lens of warm, tropical water in the equatorial Pacific. Considering that it took three years for the invader to establish a self-sustaining breeding mass in California, the advance toward BC has been swift.

On a global scale, developments like this might be compared to a shuffling of the genetic cards that comprise the world's complex inventory of biodiversity. Nobody knows what the large-scale consequences of these altered and recombined ecosystems will be. "We know that introduced species can have devastating effects," says Jamieson. He points to the Black Sea. There invading jellyfish disrupted the natural food chain, consuming the microscopic creatures that sustained more highly evolved animals during their immature stages. Populations higher up the food chain were starved out and replaced by a jellyfish population that then exploded to fill the vacant ecological niche. In effect, the genetic cards in the Black Sea deck were fundamentally rewritten, probably irreversibly. "It didn't really destroy the existing biomass," Jamieson says, "it just converted it [from many species] to jellies."

Driving these changes is the rapid growth of trans-world shipping as a global free trade zone expands. The Port of Vancouver, for example, averaged seven foreign ships a day during 1996 and, although there are regulations in place to prevent discharges of contaminated bilge or ballast water, some wonder whether there are any genuinely effective ways of preventing accidental but potentially devastating inoculations of one ecosystem with invaders from another. Greg Ruiz, a researcher at the Smithsonian Institute's environmental research centre on Chesapeake Bay—another invasion point for the green crab—warns in a recent research publication that the inevitable expansion of this ocean trade means the disruption of continental ecosystems will continue. "What we've seen over the last few decades is really an explosion in the amount of commercial traffic that is bringing ballast water to different parts of the world," he writes. "At any one time there may be tens of thousands of vessels moving around the world carrying ballast water. The effect has really been to open up a conduit for the transfer of species from one part of the world to another part of the world."

In 1997, ocean scientists from the University of Victoria and the federal government met to discuss the problem of the introduction of exotic species to BC's marine ecosystems. Minutes from their meeting expressed concerns that there was insufficient regulation of the discharge of ballast water from freighters that may have taken it on in foreign ports and that there are "a multitude of risks" inherent in the practice if captains get away with it. Ballast water samples from fifty boardings at Vancouver, Nanaimo and Prince Rupert confirmed the presence of twenty different invertebrates originating in Japan, China, Korea and the mid-Pacific.

Some of this invasion biology can actually be lethal to humans. The micro-organism responsible for cholera is believed to have hitch-hiked to Latin America in the ballast water of ships originating in China and there is evidence that the growth of the bacterium is enhanced by natural plankton populations. The hitchhiking cholera becomes virulent during plankton blooms, infects humans who eat contaminated fish and can move through freshwater systems with explosive speed. Following an epidemic in South America, studies found the cholera bacterium in the ballast of three of fourteen ships sampled. Since then a new and more virulent strain has appeared in Indian and Bangladesh while a strain resistant to chlorine, the standard disinfectant, has appeared in Peru.

"These events resulted in the death of thousands of people and with more virulent strains involved there is potential to harm many more," Tunnicliffe and McQuoid warn. "British Columbia could similarly receive this organism from ship transport and our coast does support blooms of some of the organisms previously found to enhance the growth of *Vibrio cholera*. Although our high standards of hygiene might protect us from the epidemics observed in Latin America, our sewage disposal methods (we pump a lot of untreated and minimally treated sewage directly into the ocean) would recycle the pathogen back into its optimal environment and this could result in repeated contamination of coastal fish species."

Another high-profile example of invasion biology is the human immunodeficiency virus. One of at least thirty previously unknown viruses to emerge from the tropical jungles in the last twenty years, it hitchhiked along a new highway through equatorial Africa, then rode its hosts into Europe and North America aboard ships and jet planes. More than 30 million people from the Canadian Arctic to the tropics of Asia are now infected and the virus expands its range at the rate of 6,000 new hosts per day. Viewing biological change as a function of commercial trade is a new way of conceptualizing evolutionary principles. All of which places scientists like Tunnicliffe, McQuoid, Jamieson and his equally concerned colleagues at the University of Washington, the Pacific Wildlife Research Centre in Delta and the department of zoology at the University of New Hampshire right at the cutting edge of invasion ecology studies.

Meanwhile, on North America's west coast the arrival of the green crab in substantial numbers seems to be a matter of when, not if, and these scientists are looking at the opportunity of embarking on a complete study of the invasion of a large ecosystem from the very first arrival of the alien species. While all this provides for exciting science, it has its sombre side in the practical world. The green crab clearly has the potential to simultaneously devastate commercial stocks while sharply increasing input costs for a crucial fishery. In 1995, BC waters produced 30,000 tonnes of shellfish for domestic and export markets. After salmon and roe herring, shellfish harvesting is the most important and commercially valuable fishery in the province. It is worth more than all the remaining fisheries—including cod, hake, red snapper and halibut—combined.

In the meantime, scientists have discovered the green crab capable of learning and improving its prey-handling techniques even while foraging. It is also quicker, more dexterous and has more ways of opening shells than other crustaceans, which is where the commercial threat lies. One green crab is capable of eating three oysters and up to three dozen mussels per day. It can dig out clams buried up to 15 centimetres deep in mud or sand. It is a prolific breeder. The imminent arrival of the green crab means shellfish growers in BC and Puget Sound might soon be forced to move from low-cost seabed production to a much more costly suspended system. And if experience elsewhere holds true, they will face an expensive installation of small-meshed nets to protect immature clam and oyster colonies from the voracious predator. At equal risk is the $25-million-a-year commercial crab fishery, since the green crab is an aggressive predator of juvenile Dungeness crabs, killing and eating them before they have an opportunity to grow to commercial size.

There is also concern about the invader's potential impact upon the many migratory birds which pass through the Lower Mainland and feed on a wide range of the intertidal species which inhabit BC's rich coastal estuaries. "These birds feed heavily on small crustaceans," Jamieson points out. "If they decline sharply (because of prior predation by green crabs) there could be an impact on shorebirds. It's a change in the whole ecosystem." To make things worse, the habitat in which the green crab thrives most prolifically—protected coastal ponds, lagoons, embayments and estuaries—is prime habitat for both commercial shellfish production and for migrating shorebirds.

First identified in San Francisco harbour in 1989, the European green crab probably established itself when a visiting ship pumped out ballast water which contained sufficient larvae to establish a viable breeding population. Since native species have not evolved in the presence of the invading crab, many—including most commercial species—don't have defensive strategies. And lacking a normal range of natural predators of its own, the green crab brings the potential for a population explosion. On the East Coast, for example, a single scallop pond yielded seven tonnes of green crabs over five months.

From its first sighting in San Francisco, the crab has advanced steadily, its larvae drifting northward on currents that carry it up to 8 kilometres a day. By 1997, the species had reached Coos Bay, Oregon, about 300 kilometres north of its last known colony at Humbolt Bay, California. In June of 1999, a female specimen was found at the head of Useless Inlet on Barkley Sound. Jamieson says the prevailing ocean flow patterns mean that it is now only a matter of time—a few years at the most—before the introduced species begins to invade the lucrative shellfish beds of Puget Sound and the BC coast. "This is the first time that a major predator has been introduced to our waters," says Jamieson. "It means a change in the whole coastal ecosystem."

Into the Firestorm of History

Each morning for half a century, Frank Breault has awakened to the memory of that winter day when his father strolled to Milo Grubb's barber shop and vanished off the face of the earth. It was a Saturday: February 13, 1943. What had been a sleepy frontier farming settlement that hauled its water in barrels from a spring on Bear Mountain was now the seething railhead for the world's biggest construction project.

The US Army had arrived to build a highway to Alaska and seven battalions of troops, construction crews and supplies were pouring through Dawson Creek where heavy equipment was assembled before moving up what they called the Alcan Road. In eighteen months, the population jumped from 500 people to 20,000, virtually all of them young men. Saturday nights were lively around the Dew Drop Inn and the Dawson Hotel.

Edward Nelson Breault wasn't interested in a night on the town. He was Dawson Creek's insurance agent and notary public, a long-time resident who was a happily married family man and the father of four kids: Gloria, June, Frank and Stan. In 1940, he'd been mentioned in *National Geographic* as the most northerly agent for the Wawanesa Insurance Company.

"He just went for a haircut," Frank told me, musing over a coffee at his South Dawson farm.

While his dad was in the barber's chair late Saturday afternoon, the fire alarm sounded. It was shortly after 6 p.m. Edward, like many who settle frontier communities, was a member of the volunteer fire department. In towns built of wood and without running water, any fire is a terrifying threat. As the sturdy local businessman hurried from the barber shop to answer the fire alarm, he must have been deeply worried by the chinook stirring in the western foothills.

The fire, it turned out, wasn't far away. Smoke and showers of sparks were pouring from the old livery stable, reborn as a garage for Miller Construction Co. It was approximately 6:30 p.m. Edward boosted Joe Koskick, a young truck driver, onto the long, low roof of the blacksmith shop next door and passed up a fire hose. Next to the burning garage was where Edward was last seen.

What he and the other firefighters couldn't know was that inside the building, on a truck mechanics had been hurrying to repair, was a tonne-and-a-half of US Army high explosives. When the consignment of sixty cases of dynamite and twenty cases of detonator caps exploded, the blast was so powerful that it shook the village of Rolla, 23 kilometres away, broke liquor bottles in Pouce Coupe, 10 kilometres away, and shattered windows over an area of 10 square kilometres. In the centre of Dawson Creek, the blast vaporized the garage and dug a crater as deep as a grave. The fireball incinerated some of those close to ground zero. US Army Private Norman Wasley, halfway up a ladder holding the fire hose for Joe Kosick, was killed instantly when sixty-four bolts from a box on the other side of the wall tore through his body.

Joe, the driver boosted up by Edward Breault and the last man to see him alive, still remembers with vivid clarity the faces of firefighters reflecting the flames in the pitch dark. "I was leaning with my hip against the false front. I turned my face because it was burning so hot. All I could see was Mr. Breault's round face, shining up at me out of the fire. Then it blew. Two other guys got sucked into the fire. The rest of us got blown off the building. When I came to I was 165 feet away. The guy next to me had a big wedge driven right through his adam's apple. When I opened my eyes, all I could see was a fountain of red going up over the building."

The torrent of flaming debris erupted a hundred metres above the town and descended on the six square blocks of Dawson Creek's business centre. Jennie Bowman, three blocks from the blast, saw window casements ripped from the US Engineers' mess and later described her fear to the feisty *Peace River News*. "Fire seemed to be coming down on our heads," she said.

In the immediate blast, adjacent buildings in the block were crushed like eggshells, flaming beams were smashed through other buildings. A car parked in the street remained freakishly in place but had all its tires blown off. Bill Waite, driving back from supper, was blown end over end down the street in his truck. Farther out, fronts were ripped off buildings, doors and pieces of verandas went sailing, electrical power was knocked out and streets and buildings were filled with flying glass. Joe Drushka got a face full of glass and something hit him right where he kept his old fashioned pocket watch. The blow ruptured his stomach but the old watch saved his life. Mary Basisty was on shift as a waitress at Peter Wing's café when the windows blew in. She described it as "just one great big smash...The blood was running down my uniform. It happened so fast that I wasn't sure what it was. I ran out the back door and I just kept on running." Others had the same experience. Firefighters who were knocked down by the explosion said they were stunned for a moment, then broke into involuntary flight. Pete Estebon, blown off one of the roofs, landed unscathed on his feet a block away and just kept running. Some ran like sprinters for four blocks then collapsed, exhausted, to find themselves injured with wounds they hadn't felt. Donald Waugh had no idea how he acquired his smashed arm.

The concussion knocked the town's fire engines out of commission. The firestorm that followed consumed the twenty buildings of the town's central commercial block and did damage equivalent to millions of today's dollars. Every building in the remaining six blocks sustained damage. Crouching behind wet blankets to shield themselves from the searing heat, US Army engineers used mattresses to dam roadside ditches. The puddles provided water for a desperate bucket brigade. The boiler of a Northern Alberta Railway locomotive was jury-rigged to replace the useless pumper truck. The efforts were in vain. Dawson Creek's town centre and Edward Breault had ceased to exist at the same moment.

The next night, with martial law in place and soldiers patrolling the streets, witnesses at nearby Pouce Coupe said the horizon over Dawson Creek still glowed a dull red. Meanwhile, hundreds of burn cases, lacerations caused by flying glass, concussions and contusions caused by heavy debris had descended on tiny St. Joseph's Hospital. A doctor and four nurses worked feverishly by candlelight while the injured overflowed into corridors and waiting rooms. Two nurses, Mrs. William Koeniger of Winnipeg and Doris Davidson of Toronto, volunteered for a dangerous night flight from Fort St. John to bring relief. The plane was already aloft when someone realized there were no runway lights. Men dashed to the air strip and set out highway flares but in the darkness and confusion some were set too close to the end of the strip. On its descent, the aircraft clipped telephone wires but managed to get down without serious damage. The two relief nurses rushed to St. Joseph's where they worked without break for the next twenty-four hours. Seventeen of the most gravely injured survivors were evacuated by US Army plane for treatment at Edmonton hospitals.

Back in the chaos of the townsite, families and friends hunted for the lost among army stretchers, hospital wards and the boarded-up building that served as a morgue. For those without family, workmates took charge. "The next day at noon the boys found me on a stretcher in the army camp," recalls Joe Kosick. "Andy Janes, my part-time driver, he yelled 'There he is!' and he gave me a kiss I'll never forget." Miraculously, only five—possibly six—people are believed to have been killed in the explosion. One of them is thought to have been nine-year-old Johnny Frederickson, the son of a water hauler who sneaked down to watch the blaze, another is thought to have been Edward Breault. The blast occurred while thousands of young soldiers were still in the army mess on the outskirts of town. Had the explosion happened thirty minutes later, Saturday night crowds would have thronged the street outside the garage and the Dawson Hotel beer parlour, which vanished in the firestorm, would have been jammed with off-duty troops and construction workers.

Surprisingly, the explosion doesn't rate a line in the *World Almanac*'s chronicle of notable Canadian events for 1943. It's not mentioned in the *Canadian Encyclopedia* entry under Dawson Creek and it doesn't even make the *Canadian Global Almanac*'s list of Canadian disasters. How did such a momentous wartime incident come to have such minuscule impact on Canadian consciousness? Some of the factors that resulted in confusion over the ultimate number of casualties undoubtedly contributed.

Japanese and German military expansion was at its zenith and although the tide was about to turn, nobody knew it. Across the Atlantic, the Canadian Army was regrouping from the humiliation at Dieppe six months before and the fate of the war hinged on the titanic struggle for Stalingrad. In western Canada, attention was focussed on the battle of the Solomons, where the US Navy had just lost its heavy cruiser *Chicago* and twenty-two warplanes. The Alcan Highway construction project, completed only ten weeks before, was considered of highest strategic priority and the subject of heavy wartime censorship.

Even in Dawson Creek, authorities had difficulty establishing how many people were actually missing after the blast because so many of the contract civilian crews were transient, unattached men without dependents. To make things worse, 6,000 ration books were destroyed in the fire. St. Joseph's Hospital was unable to record the names of all the injured because, as the Sister Superior later confirmed, the medical crisis quickly overwhelmed any administrative considerations. "We were so very busy. We just did what we humanly could. After it was nearly all over we sat down to take down the names of the patients who were still with us, and the names of some of those who had left and we remembered."

After a long investigation, the US Army accepted responsibility for the disaster and Amelia Breault was paid a lump sum settlement. She took her children to live with their grandfather on the South Dawson homestead he settled in 1917. The old man had freighted-in the second load of groceries for the Dawson Creek Co-op, founded with the $600 collected when each farmer chipped in $25. Ironically, the Co-op was the only building to survive the blast unscathed. Dominic Torio shaped his grandson into a skilled north country farmer, but he never replaced the father who vanished that Saturday afternoon so long ago.

On the fiftieth anniversary a commemorative mass was held in Dawson Creek. Joe Kosick, the last man to see Edward Breault on this earth, and other survivors prayed for the dead friends of their youth and gave thanks for the gift of the rest of their lives. But every morning, Frank Breault still wakes to the incandescent memory of the father who walked out of a little boy's life and into the firestorm of forgotten history.

A Dry Wind in Eden

The arid heartland of south central British Columbia is empty and austere. Around Kamloops, defined by talus slopes and painfully blue skies, the sunburnt grassland is punctuated by drifts of pine. The setting is startlingly park-like, announcing itself to even the untutored eye as near-perfect horse country. Rain comes seldom here—less than eight inches of precipitation a year, less even than New Mexico—and summer temperatures normally soar into the high thirties. Far below these parched tablelands, intense and inaccessible as a mirage, churns the huge gun-metal sheen of the Thompson River. Scattered along its banks are the small green oases of irrigation farms. Robot sprinklers, spindly assemblages of wheels and pipes in perpetual bondage to a central water source, chuff tirelessly around lush circles. They signal the technological hubris inherent in any attempt to satisfy the immense thirst of the countryside. The life-giving mist they discharge carries away on furnace winds. Much of the water evaporates before it reaches the ground. Here and there, at the neglected margins of two-crop alfalfa fields, a few twisted trees struggle to survive.

Across this crumpled cordilleran rainshadow snakes the

Trans-Canada Highway. It reaches westward from Kamloops toward Cache Creek and the oven that is Hell's Gate Canyon. Along its course, the heat shimmers that accompany the fierce days of high summer spawn frequent dust devils. They materialize, fitfully toss litter and dead grass in an impish, elemental roadside dance, then just as suddenly wink out of existence. Above this ribbon of blacktop, twisting mile after monotonous mile across the dry gulches and eroded benchlands, attended only by sage brush and the region's small indigenous cactus, a peculiar trail of debris parallels the highway. The long stain on the hillsides is the wreckage of an abandoned irrigation flume. Dessicated by the climate for nearly a century, now disintegrating with the assistance of exhaust emissions from countless passing cars, this decomposing wood frames a doorway into the guttering nightmares of my grandfather's generation, a collective experience now almost vanished from living memory. The splintered flume to Walhachin represents one more decaying monument to the twilight of a ruined civilization.

The dry air of this desert country resonates with ghosts, even in the organic music of common place names. Black Canyon evokes Chief Trader Stuart Black, shot down in an Indian blood feud in 1841. Ussher Creek recalls the policeman killed by the MacLean Gang in 1879. Monte Creek is where Bill Miner robbed his last CPR train in 1906. Ingram's Grave is the golden knoll from which early rancher Henry Ingram has an eternal prospect of the sere grasslands that roll away to the south. And there is Walhachin, the settlement that vanished into thin air. Even the rotting flume that was to make the desert bloom evokes a ghostly echo. It descends from the Deadman River, itself named for Pierre Chivrette of the Fur Brigades, killed in a quarrel over a campsite in 1817.

But the truth is that I no longer care to learn the facts about Walhachin, the story of its founding near Charles Penney's homestead by American Charles E. Barnes or how it passed into the hands of the Marquess of Anglesy. I was hungry to know such things once, a skinny kid discovering history and literature, keeping notebooks full of details, gathering stories at the old folk's home where I delivered papers, tapping even then into the seductive power of finding things out, a passion that would take me first to journalism and then beyond it. At the brink of adolescence, perhaps the world is always encountered as a coded message, a secret rich with the possibilities of decipherment. By fifty, a growing awareness of how little can be deciphered from facts begins to take root. Penny's Flats was Penney's pre-emption but went into the history books as Pennie's Ranch. For all I know of the mystery of Walhachin today, after almost forty years of bearing witness to its strangeness, the truth emerges less from facts like these than from folklore lovingly burnished into legend.

A short drive west of Kamloops, the bustling urban descendant of David Stuart's swindling of the Shuswaps—he traded $125 in tobacco and cotton for $12,000 in beaver pelts—Walhachin is now little more than a curious name on the maps. And like so much of the past we invent and reinvent for ourselves in Western Canada, even the name serves as one more small duplicity of British imperialism. The blocky European spelling is merely a best guess at the liquid syllables of the Salish tongue—was it Walhassen or Wal'a-sn?—later modified to Walhachin to make commercial propaganda. Thus do characters from the dead language of one fallen empire reach around the globe on behalf of another to colonize and shape a pre-literate culture more ancient than both. These precisely chiselled Roman letters serve to anglicize an ornate bit of marketing that gilds the lovely Indian idiom, honest and plain as the place itself. "At the edge," was what the Thompson Indians called the place, doubtless referring to the steep 335-metre descent to this mighty tributary of the Fraser. "Round stones," concluded a later European translator, presumably in reference to the huge water-polished boulders below. "An Indian word signifying an abundance of food products from the earth" claimed the nineteenth century land promoters, looting the language of the Ghost Dance for a name with marketing appeal.

The story of Walhachin now seems curiously mingled with the idea of the Ghost Circle of the Salishan peoples. It too is a place populated by ghosts, indeed, became a ghost itself, haunting the flickering memories of old men and women. To remember Walhachin is to set a place at the table for the absent, to see our own imperial history drawn somehow into the Indian myth, colonized by it in return, rendered curiously similar and returned to us. At Walhachin, the past is strong medicine. Indeed, the Thompson Indians' Chief of the Dead, a powerful transformation figure, became oddly superimposed upon the wrathful Jehovah of the Christian missionaries who came to dethrone the resident spirits. At what was once the Kamloops Indian residential school, now a museum, one can still see bizarre photos of grinning priests celebrating St. Patrick's Day by dressing up solemn Indian children as leprechauns. Yet to visit what once was Walhachin, to stand in the amber light of evening with purple shadows spilling from the coulees and mud swallows darting above the muscular coils of the great river is to sense that only the gods of commerce and the white man's burden have proved transient.

The stunted, sun-blasted scrub at the field margins is what remains of Eden. These are the age-blackened remnants of 16,000 trees that once tossed and breathed, filling the hot valley with a sound not heard since the inland seas retreated eons before. Once you've heard that silver-bellied rustling of leaves on the wind, a susurration like the faint whisper of surf on distant coasts, the sound clicks forever into the templates of memory. In the soft summer nights of my youth, the darkest nights of the Cold War, I would slip out of the old house that creaked and groaned like a ship moored on the heaving green swells of the orchards. I'd lie awake looking into starry skies for the transit of Russian satellites, listening to the restless movement of the living fruit trees around me. Even now, almost half a century later, starting up out of sleep, I sometimes hear that sound in the night, just as the vanished souls of Walhachin must have heard it in the troubled mutter of barrage and counter-barrage. Do any survivors dream it still in the quiet cells of their nursing homes?

Those tossing trees once perfumed the high country with the scent of peach blossoms, pear and apricot, Winesaps and Wagoners and Jonathons and Cox's Orange Pippins. Walhachin was to be the reinvention of the lost biblical garden, a vision of scorched fields made to flower in a bone-dry wilderness. There is disagreement among historians regarding the truth of what proved to be the worm at the heart of that vision. Some say the settlement went down in the simple ebb and flow of commerce, a bad investment poorly conceived, commodities too far from market, another of the utopian adventures that stud the West with ghost town ruins. But my story says Eden was laid waste during Canada's bloody birth as the industrial nation we now inhabit.

The legend of its fall from grace was told to me as a boy listening to old men with stubble on their faces, old men patiently awaiting their own final ticket out of the green stuccoed senior's home that came last on my paper route. Bony knuckles clicking under transparent skin, they'd clutch my hand and tell me of the slain heroes. One gave me his medal, a luminous disc that showed a naked man with a sword riding a horse over a field of skulls and bones. Another told me how, in the dressing stations, they'd divide the casualties by thirds—one third to walk to the next station, one third to the surgeons, one third to the dying tent—and how the dying would sometimes moan as they were taken out: "No! No! It's the wrong place!" This was how I learned the theory of medical triage, although I wanted only to get out of the dark room with its sickly sweet odour of age, to fly on my CCM bike to the beach where I'd spread my towel on sand too hot for the skin and lie there with my sweating Hi-Spot, hoping for a glimpse of the suddenly changed girls from grade seven, that other mystery waiting to be decoded.

I told these old men about the strange wooden flumes I saw from the window of my father's green Studebaker on Sunday afternoon expeditions. My father was a man who loved finding things out, a man devoted to revelations, discoveries and disclosures, the kind of man whose idea of fun was to take his kids through the Rockies when the Yellowhead Pass was a dirt track, just to see if it could be done. I loved to ride up in the cockpit of that bullet-nosed car with its divided windshield and its front end like what I imagined a Spitfire fuselage must look like. My brothers and I would sight the on-coming traffic along the hood ornament, shoot them down, jeering as the hump-backed Oldsmobiles and Pontiacs disappeared into the slipstream. My father would look angry and upset but said nothing. Boys at the brink of teenage rebellion seldom ask their father what troubles him. It was years before I learned how he spent his own war in prison as a conscientious objector: a pacifist from a family with a military tradition, his own father a career veteran of the Imperial Army wounded at Gallipoli. His cousin Jack interceded for him, announcing "That's why I'm there, so lads like him can make a choice"—his beloved Cousin Jack who was slain in the fierce armoured battles around Caen in 1944.

In the old folk's home, my fading elders told me the rotting flumes were all that remained of Walhachin, a town murdered by the First World War, a flower cut down in a single desperate cavalry charge against mechanized weapons. The War To End War, they told me, changed everything, was the end of everything, the death of innocence, the birth of the infinite evil that twenty years later would define our century. I choose now to believe the truth of their legend, whatever the arguments of paper-shuffling academics regarding the facts of economic predetermination. I've seen some of that truth with my own eyes. The old man with the strangely mottled skin and unkempt hair, his eyes darting wildly, gabbling partial words at the frightened kids on the dusty school grounds. Not to worry, we were reassured by our white-haired English teacher, a woman who seemed even older than the scarecrow man. "Shell shock," she told us, interrupting our bowdlerized *Romeo and Juliet* for a brief lesson in social history. She explained the paralyzing artillery barrages of the war before the war my own father refused to talk about—barrages so horribly intense, she said, that they shook the wits right out of men. Some were still in hospital forty years later, trapped in the memory of battles they could not escape. This seemed unbelievable. And then, years later, my long-dead teacher's story surged into remembrance with vivid, visceral intensity when I stumbled across the newspaper obituary for Jim Wilson, the Canadian boy who spent sixty-one years in the psychiatric wards and died at the age of 96, still trying to find his way out of the mud-filled trenches of Passchendaele.

There had been twenty-six men in his unit of the Toronto Scottish when German shrapnel killed or mortally wounded twenty-four of them. Orders to fall back came up, but he refused to abandon his brother John. He went back among the dead and dying, wiping mud and blood and brains from the faces. He found his brother, slit open from hip to shoulder but alive and carried him in through the mud and slaughter to a forward field hospital. "Here," he told the medics as he put the wounded man on the operating table, "this is my brother. Look after him. You have to save him." Then he stumbled back out into the night of carnage and chaos and collapsed. When he woke hours later it was on a pile of stiffening corpses. Jim Wilson and his brother were the only survivors. He was, in the words of the day, never quite right in the head again.

But back in grade seven the only war I knew was the one John Wayne showed us in the two-bit black-and-white Saturday matinees. Even then I had noticed that The Duke didn't sweat. When my father finally did speak of such things, it was to tell me of a recurring image from his own war—he served as a first aid medic during the Blitz—the image of a little girl lying in a Coventry gutter, long hair streaming out in the black rain, killed in the blast from a German bomb. He told me how the dead will sometimes groan as the process of internal decomposition expands gases. How rigor mortis comes and goes. How rows of corpses must have tickets neatly tied to them in the standardized rites of identification. How when his turn to be called up came, he refused to turn other men and women and children into offal. "That's war," he said of the dead child. "Not your Grandad's medals."

Walhachin was to be a metaphor for Eden, a vision of pastoral harmony conceived before anyone could imagine destruction on such a scale: whole cities razed, great cultures debauched, entire peoples slated for extermination. The settlement in the western desert was the culmination of the dreams of an American engineer and British aristocrats still caught up in the follies of colonial expansionism and the romance of empire. It was to be a model community, a cornucopia, a marriage of Yankee know-how and Imperial noblesse oblige. It would have homes located to exploit the view, a monorail, an elegant hotel offering excellent cuisine, carriage tours of the orchards, a carefree life of parasols and polo.

To some extent, the Walhachin scheme was carried forward by the peculiar momentum that brought a flood of British settlers to the Okanagan Valley and the Kamloops region between 1890 and 1914. Some came with great expectations and a sense of biblical mission—"Get thee out of thy country, and from thy kindred, and from thy father's house," says the injunction from Genesis on a window in Penticton's Memorial Chapel dedicated to early settler Tom Ellis. Some came in search of adventure, the generation that missed the Cariboo and Klondike gold rushes. Some came to hide indiscretions from the wrath of a rigidly stratified society. Some were sent in an effort to put as much distance as possible between themselves and their proper Victorian families. "The country was full of the queerest people you ever met in your life," recalled Bob Gamman in an interview for the BC Provincial Archives. "They all had a history behind them, you know—they all had a history. They were wealthy boys and remittance men, lots of them. But they'd all had experience. They'd hunted in Africa. They'd been to India. Why they ever came to Canada, I don't know. But they were real men, they really were."

One of them became Dorothea Walker's husband. She, too, reflected for the archives on a young man bound for the clergy, schooled in Latin and Greek at Oxford, who at nineteen defied his parents. He would not become a minister. "'Well,' his father said, 'it'll have to be the colonies'," she recalled. "And so, the result was, he was sent out here as a pupil to learn farming at $500 a year. He arrived in September in a tweed Norfolk coat and knickerbockers, you know, and woollen stockings and a tweed cap." Mud-pups, the locals called them.

Adventurer, fortune hunter or remittance man, they all had one thing in common: everybody was fruit mad. The orchard boom had begun when Lord Aberdeen purchased the Coldstream Ranch near Vernon in 1891 and established the first commercial orchard in BC's interior. Earl Grey, Governor General of Canada, expressed the mania in an address to the Royal Agricultural Society to open the New Westminster Exhibition in 1910:

> Fruit growing in your province has acquired the distinction of being a beautiful art as well as a most profitable industry. After a maximum of five years I understand the settler may look forward with reasonable certainty to a net income of from $100 to $150 per acre, after all expenses of cultivation have been paid.
>
> Gentlemen, here is a state of things which appears to offer the opportunity of living under such ideal conditions as

> struggling humanity has only succeeded in reaching in one or two of the most favoured spots upon the earth. There are thousands of families living in England today, families of refinement, culture and distinction, families such as you would welcome among you with both arms, who would be only too glad to come out and occupy a log hut on five acres of a pear or apple orchard in full bearing, if they could do so at a reasonable cost.

The partners at Walhachin easily obtained financial backing in Britain and began courting the right kind of settlers among the English gentry and would-be gentry. In 1908 they formally founded the settlement on 2,000 hectares that spanned the Thompson River at the flats near Penney's original homestead. It was to offer a country squire's genteel lifestyle in which the leisurely tending of orchards would yield bountiful, well-paying crops. There was only one problem. The dry land would not support fruit trees and the river was so far below the benches that pumping water for irrigation was uneconomic. No matter. With absolute Edwardian confidence in the powers of engineering to overcome natural adversity, the owners began immediate construction of 20 kilometres of wooden flume to bring water from Deadman River. They planted their vast orchard and prepared to reap the bounty. By 1910, fifty-six settlers had come to Walhachin, but not to Earl Grey's log huts. More than a dozen families were ensconsed in graceful new houses with large verandas. French doors opened to the garden and hand-carved woodwork ornamented the exteriors. Those who were still waiting for their houses to be completed took up residence in the Walhachin Hotel, says Kamloops historian Joan Weir, who details the settlement's origins in a slim but meticulous monograph published in 1984. Indeed, many of these upper-crust pioneers made a practice of skipping the winters to catch up on London society, to investigate prospective brides or to take the grand tour of France and Italy.

In short order, the community had a restaurant, a laundry, livery stables, competing newspapers, a bakery, ladies and gentlemen's haberdasheries, real estate and insurance offices. Gordon Muriel Flowerdew, one of the dashing young English bachelors who increasingly populated the countryside, ran the butcher shop and general store. His sister, Miss Eleanor Flowerdew, ran the Walhachin Hotel with its cool verandas and crisp white linen. The cuisine was of the highest order and guests were expected to dress for dinner in full evening attire. "The large dining room overlooks the orchards and from the open balcony on the north side a good view is obtained of the orchards and of the Thompson River," said an enthusiastic review in the *Ashcroft Journal*. "Spacious billiard, card and ladies' and gentlemen's sitting rooms occupy the east wing. Hot and cold water, gas and all comforts have had careful attention." There was a football club and a cricket side and golf and afternoon tea and musical evenings and many formal balls with top hats, tails and white kid gloves. Residents even rode to the hounds, although a coyote had to be substituted for a fox.

At this peculiar outpost of the British Imperium, military traditions came with the polo ponies. Lieutenant Colonel J.D. Willson, who served with the Mounted Rifles in the Boer War just a few years before the founding of Walhachin, had this to say of Canada's west: "Perhaps in no country in the world was there finer recruiting ground for light cavalry, for though the population was sparse and widely scattered, most men rode and a considerable portion of them were horse-owners and daily in the saddle...Our horses would have compared favourably with those of any light cavalry in the world for speed, strength and beauty, and even superior to most in toughness. Our ranks were filled with every class of the West, generally drawn from the farmers and stockmen, of a physique rarely equalled by any regiment I have ever seen."

Gordon Flowerdew exemplified this ideal. A Walhachin Company of the British Columbia Horse had been raised in the summer of 1911 and the following year he was noted for setting records in both shooting and steeplechasing at the cavalry training school in Vernon. He was particularly mentioned for his performances in the Victoria Cross Race. The summer his world came to an end was tawny and tranquil, infused with a serenity we may never know again. When the Kaiser invaded Belgium and Britain declared war on Germany on August 4, 1914, Flowerdew was among the first to volunteer. Elsie Turnbull, writing in *Pioneer Days in British Columbia*, says that the volunteers soon included virtually the entire male population of Walhachin and that 97 of the 107 men in the village had joined the army to defend the British Empire in France.

Given the well-deserved cynicism of our own times, it is difficult to imagine the excitement—even joy—with which the outbreak of war was greeted across the West by young men who saw an opportunity to shed bucolic routine for some higher purpose. The war was to be like the game of polo but played for higher stakes. Flowerdew admitted his boyhood dream was to win the Victoria Cross, although Weir writes that he later confided to an officer, with suitably becoming modesty, that "I shall never be brave enough to win it. Valour has reached such a standard that you have to be dead before you win the VC."

Older, wiser souls like novelist Henry James, whose own youth had borne witness to the American Civil War, had a deep foreboding about what lay ahead in the collision of industrial titans. "Black and hideous to me is the tragedy that gathers, and I'm sick beyond cure to have lived to see it," he wrote in a prescient letter from England to Rhoda Broughton dated August 10, 1914.

> You and I, the ornament of our generation, should have been spared this wreck of our beliefs that through the long years we had seen civilizations grow and the worst become impossible. The tide that bore us along was then all the while moving to *this* as its grand Niagara—yet what a blessing we

didn't know it. It seems to me to undo everything, everything that was ours, in the most horrible retroactive way—but I avert my face from the monstrous scene...The country and the season here are of a beauty of peace, and liveliness of light, and summer grace, that make it inconceivable that just across the Channel, blue as paint today, the fields of France and Belgium are being, or are about to be, given up to unthinkable massacre and misery.

All through British Columbia's boundary country, throughout the Okanagan, across Vancouver Island and out into the foothills and plains, spurred in no small part by the legendary North West Mounted Police and the exploits of the Mounted Rifles in South Africa, young men were clamouring to enlist in regiments with Kiplingesque names: Tuxford's Dandys from Vernon and Merritt; the Rocky Mountain Rangers; the Alberta Dragoons; the Kootenay Borderers; Lord Strathcona's Horse and the British Columbia Horse.

The first troops bound for the front left Victoria for Valcartier mobilization camp in Quebec on August 26, 1914. The train stopped as it wound through the high summer valleys where the apples were sweet and heavy on the trees, picking up the men from interior regiments anxious to abandon the fall harvest. One vivid description of a troop train carrying 519 men from the 225th Kootenay Battalion in 1915 was typical. It came over the Kettle Valley Railway, arriving at Penticton at 6:30 p.m. after being delayed for five hours by washouts at Midway and Rock Creek. "It seemed as if the entire population of the town had turned out to see the soldiers," reported the *Penticton Herald*. "The long train of 11 or 12 cars was jammed full of khaki-clad boys. Heads protruded from every window and the travelling Kootenay volunteers took a decided interest in what they saw here."

Half an hour later, formed into columns on the wharf, the troops marched company by company past the fruit-packing sheds to the gleaming white paddle steamer *Sicamous*. As they travelled down the glass-smooth lake among the reflections of puffy white clouds in an azure sky, the soldiers enjoyed 500 pounds of cherries, a special harvest that women had picked in the cool, green, irrigated orchards and brought down to the dock to send the boys on their way to glory. I imagine them biting into the crisp flesh of those ripe cherries after a long, hot day, the laughter and the spurts of dark red juice that splattered on the white paint where they spit their pits over the side.

"Little Walhachin, with a total population of under 150, sent 43 skookum men to fight for King and Empire," marvelled C.L. Flick in his account of the early mobilization of the Canadian militia. The new-founded University of British Columbia had one-fifth of its student body enlist. The sparsely populated area around Kamloops, which included Walhachin, had 4,000 men in arms. By war's end, BC, with barely 450,000 people, would have 55,570 in uniform.

This heavy enlistment was typical of the patriotic fervour which accompanied the early days of the Great War. "It was like an oil boom, or the subdivision of a new town site; offices sprang into existence, crowds waited at the doors, and, inside, the investors made their deposits. This was no idle punting in oil shares or mining stock; it was a solid investment in flesh and blood," wrote Herbert Rae.

> Pete Sornson from Fort Charles was deficient of a hand and kept the stump carefully hidden behind his back, until told to spread his fingers out. That ditched him. Andie Mack from Squamish, with a wooden leg, the result of a badly primed dynamite cartridge, kept the fact concealed until told to take his trousers down. There were lumbermen and railwaymen, prospectors, surveyors, bankers, brokers, stokers, teamsters, carpenters and schoolmasters. Many had not seen a city for months, and they were frequently drunk; but the material!

In August of 1914, the men of Medicine Hat joined the stampede to volunteer for the expeditionary force in France. Like the riders of Walhachin, they too were High Plains wranglers, among the best horsemen in the British Empire, world polo champions and bronc busters, soldiers of fortune who had taken a liking to the cottonwood coulees of Palliser's Triangle and stayed behind while the railhead pushed west into the Crowsnest Pass and the coal fields. Today you can still turn up the tattered IOUs they left pinned to the wall in places like the billiards room of the Cypress Club as they dashed out, rushing to be first from the mark in that fateful adventure. Four years later, one out of every five inhabitants of Medicine Hat was dead or missing in a soldier's grave. The list of names assembled for the yet-to-be-built cenotaph plunged a town once known for its gaiety into a deep and enduring mourning. The grief lingers today, a brassy aftertaste to community life in Medicine Hat the better part of a century later.

Once, writing in the dry shorthand of a newspaper article about Medicine Hat's memorial park, I was later rewarded with a powerful, vivid letter. It bore the barely legible hand of a woman whose name I keep private, as she had kept her own grief private for a lifetime. She told me how her narrowing memory still rang with the sound of her lover's polished boots booming on the veranda, the groan of the whitewashed gate closing forever on her kitchen weeping and her youthful dreams. When I read that letter again, I hear the gates closing on Walhachin, all ninety-seven of them, and the weeping of women like Miss Flowerdew, walking on the windy headlands above the timeless blue river so far below. In her war, many of the living came to envy the dead, certainly those who lingered on in mutilated bodies and maimed minds—like my schoolground scarecrow man—condemned to frighten children, objects of grateful pity at first but soon reduced to irritating curiosities. Of the 43,202 British Columbians who served overseas, almost 20,000 would become casualties. From 1919 to 1933, the number of disability pensions in Canada would swell by 75 percent.

Unlike those who became names carved into gleaming white gravestones, the living victims, those buried alive like Jim Wilson, are listed only in the mouldering records of the national archives, papers so marginal to history they have been removed from the main stacks for storage in a distant warehouse. One can still pay the small tribute of a walk down from the Parliament Buildings to the archives, take the trouble to disinter the call numbers of obscure files, wait the two days it takes to retrieve the records and immerse oneself in the endless 1916 lists of Canadian casualties in British hospitals.

More than eighty years later, the chronology of suffering is still sufficient to shock. At first the hospitals themselves filled up, then public buildings were commandeered, then schools, then the stately homes of the English aristocracy. The files document the confused logistics of special hospitals for the legions of the blind, for the gassed, for the burned, the legless, the armless, the faceless. And yes, the bloodless memoranda reveal even the personal trauma of military bureaucrats grappling with the moral implications of men whose facial disfigurements were so grotesque as to prohibit their release into public life. What was the future, in an era before reconstructive surgery, of a man whose lower jaw had been torn completely away? And what was the nature of a culture which preferred that he should die rather than linger to offend our eye?

When I opened the first box of water-stained cables and smeared carbon copies, I reeled from the stench of death that rose out of the stiff cardboard boxes. It was, I know, only an active imagination associating the musty odour of slowly decomposing paper with the smell of trench warfare. But I felt as though I had opened a grave from which the ghostly scent of the Great War had emerged to hover above the names of its forgotten victims. It seemed an appropriate dopplegänger, certainly more true in its way than the illuminated script and gleaming parchment of the official Roll of Honour which graces the Peace Tower in Ottawa.

Each day that stainless page is turned by an immaculate white-gloved hand to present the public with the names of those who had the good taste to be slain. The names of those who managed only to suffer and survive, to endure the embarrassments of missing jaws and faces, the shell-shock cases, they remain buried in the banker's boxes at the Renfrew storage facility. Or in the yellowing medical studies that still surface in used book stores, catalogues of the infinite varieties of battle neurosis: men condemned to cower in a protective crouch for the rest of their lives; a man who could only find sleep in a hole in the ground.

Battalions raised in western Canada commonly provided the cannon fodder for the War To End War. They were crammed into the front lines at a rate of 1,000 men for every 800 yards of trench and their lives were spent at the rate of ten men for every yard of ground gained. "I hate this murderous business," wrote Talbot Papineau in one letter from the front. "I have bound up a man without a face. I have tied a man's foot to his knee while he told me to save his leg and knew nothing of the few helpless shreds that remained. He afterwards died. I have stood by the body of a man bent backward over a shattered tree while the blood dripped from his gaping head. I have seen a man apparently uninjured die from the shock of an explosion as his elbow touched mine. Never shall I shoot duck again or draw a speckled trout to gasp in my basket—I would not wish to see the death of a spider." Some estimates hold that 60 percent of the actual casualties to the Canadian Army came from west of the Great Lakes. Perhaps this is not surprising considering the West's enthusiasm for the grand adventure and the correspondingly grisly casualty rates suffered by Canadian regiments in the front line.

The civilization from which the volunteers' romantic ideas sprang was snuffed out forever in a cloud of poison gas in the Ypres salient in 1915. Poet and classicist Guy Davenport argues persuasively that the twentieth century itself was stillborn in the mud of Flanders, that our own age actually limps on through a cacophonous interruption between cultural epochs, a castrated, shell-shocked culture incapable of self-realization. The sudden change in the character of newspaper headlines in 1915 lends weight to Davenport's argument. Ebullience gave way to grim reports of gallant young westerners dying not in glory but, in the angry words of the *Calgary Herald*, "mowed down like sheep" in a battle where British generals marched them into the machine guns in neat, straight lines. The Canadians had died with great courage and in perfect alignment. Their staff commander, watching with binoculars, wanted to know why the damned colonials lay down and refused to advance. "Because they're dead, sir," a junior officer volunteered.

Four days apart at the end of October 1917, the South Saskatchewan and Edmonton regiments advanced into a sea of liquid mud. The wounded sank into the slime and drowned, lying submerged until the gases of putrefaction squeezed the bloated corpses back to the surface weeks later. Shell holes filled with water and crusted over with a thick, lethal scum. It bubbled and seethed, occasionally belching out mustard gas to burn or blind the unlucky soldier struggling past through the mud. The formal history of the Edmonton regiment is most powerful in its cruel understatement: "Under the cold stars it seemed to many that they were alone in a dead world, a world remote from all the warmth of living, a world in which primal forces strove insensately."

Dawn and dusk were greeted with the monstrous uproar of barrages and counterbarrages down hundreds of miles of front. At night, the ghastly landscape was illuminated with starshells. The sputtering red ghosts of Very flares signalled sorties and ambushes and always, pulsing like sheet lightning along the lines, were the orange and magnesium-white of exploding shells. For the men of the Edmonton and South Saskatchewan regiments the sojourn at the front was brief. Almost eight out of ten soldiers were killed or wounded in the first day of battle. In the BC regiments, where one in ten of the total provincial population was serving, almost half were to be killed or wounded by the end of the war.

In 1915, a Sergeant Major Hill writes friends in Trail to say he'd visited the 7th Battalion at the front: "I am very sorry to inform you that I found very little except the name left. C.P Jones is the only one of the Trail boys. At the battle of Ypres the 7th Battalion lost, killed and prisoners, about 700, including 13 officers, and on returning to Billets, could not muster 100 men. Most of the Trail boys are killed, wounded or missing...Young Reese went through everything until yesterday, when he went to the hospital with a wrenched leg. C.P. Jones is well except a touch of nerves, and I say the man who comes through an engagement like the one we have come through with nothing more than a 'touch of nerves' is lucky."

Then the lucky C.P. Jones writes the *Trail Times*: "Out of the Trail boys there is only A.E. Hall and myself left; Simmons, Merry and H. Kirby are missing, and so far I have failed to locate them. In fact the Kootenay boys have suffered badly. I don't think I could find above 20 now, out of the 120 that joined this battalion."

A Sergeant Jackson writes, too: "We were in the trenches seven days and it seemed seven weeks, as the Germans were shelling us pretty accurately. We advanced under a hail of bullets and shrapnel. Our company lost heavily. I lost all my section except two men. Dwyer got injured by shrapnel, but is doing well now. Morgan had his legs broken by a shell striking his dugout; a sergeant from Fernie who was with him at the time, was killed. I was buried alive once but the boys soon had me out. Out of four officers with us we have lost three...All of us are pretty well shaken up with the strain."

It is difficult now to grasp the depressing fatalism, the sense of a vast, irresistible doom approaching that would soon grind everything to powder. I got my own first glimpse of it as a boy at mid-century, accompanying my parents on a holiday to the long white beaches which arc down the west coast of Vancouver Island. There, in a water-stained cottage nestled among towering sand dunes, the pictures on the walls were of stiff young men who had flocked to the colours in France and found themselves, in the succinct words of Ezra Pound, "eye-deep in hell." Throughout the nights, the sea hissed and whispered, making the familiar sounds of apple orchards in the wind. By candlelight, I read old magazines from the closet reporting battles at places with strange names—Polygon Wood, the Salient, Passchendaele and Amiens. Not until decades later, after a long day's journey through the microfilmed pages of 1915 newspapers, did I fully grasp the sense of doom and futility that fell across a generation of survivors. There is a moment in the public record when even the propaganda and jingoism turn to sepia and the spirited names of the regiments are replaced by numbers as the war itself is changed by the statistics of attrition. The 1st Battalion gives way to the 31st, the 50th, the 222nd. Even the most grandiloquent headlines wilt into hollow nonsense.

The first Canadian airman killed in action was Lieutenant Stanley Winther Caws of Edmonton, "a grand character, the life and soul of our little party." Aged 36, he had been a horseman in the Boer War and enlisted with the 19th Alberta Dragoons before transferring to the cavalry of the air. By 1918, 40 percent of the Royal Flying Corps was comprised of Canadians—one reason the British would not permit formation of a Canadian air force to match the First Canadian Division fighting below. To do so would have dismembered their own air corps. This would not concern Lieutenant Caws. He and his aging reflexes lasted only four months after graduating as a pilot in 1915. As the reels of microfilm flickered onward into 1916 and 1917, the daily catalogue of dead and wounded got longer and longer. Soon the British Expeditionary Force casualty list was averaging 35,000 names a month. At Passchendaele, the South Saskatchewan Regiment lost 70 percent of its men in a single afternoon. Forever afterward it was to carry the nickname of the Suicide Battalion. Edmonton's 49th Battalion lost 75 percent of its men a few days later. On November 21, 1917, the list of casualties took a whole page of the *Vancouver World*. Two days later, the new list took another full page. Then, abruptly, the long lists of names vanished from the pages of the newspapers—the deaths of local heroes became a state secret, shared only by the generals and the bereaved.

Somehow all this dying and grief seeped into the marrow of the West, burrowed into our collective literary imagination. Today, as World War I recedes from living memory, the process of its transformation into myth intensifies my own need to understand the enormity of this hinge in Canadian history. Somewhere, somehow, in the endless churning under of Kamloops cowboys and Walhachin gentry, Canada claimed its nationhood. It was there that the newly sovereign nation abandoned its agrarian roots and entered the industrial revolution. Returning soldiers, themselves profoundly changed by war, found a country transformed almost beyond recognition. "The place we had left off wasn't there anymore" wrote Norman James in *The Autobiography of a Nobody*, published thirty years later. And it was during this process of transformation that many of our still-festering wounds were inflicted—the wounds of class and language, labour and capital, urban and rural, French and English, East and West.

For Canada, 1917 was a year of crisis, both at the front and at home. In the view of historian John Herd Thompson it was "the most critical year since Confederation." Russia was sidelined by its revolution and fear of the Bolsheviks swept the west; the French army was on the brink of mutiny; the Italian army had collapsed at Caporetto. At home, "aliens" were interned in concentration camps while proponents of the One Big Union denounced a war in which it said the working classes of Germany and Canada made the bayonets which conscripted workers were then ordered to stick into each other while the munitions tycoons profited. And this seemed true on Vancouver Island where a Krupp-designed mill produced wadding for Royal Navy guns used to sink Kriegsmarine warships. Even the capture of Vimy Ridge, the greatest allied victory of the war, had been accomplished only at the price of 4,000 Canadian lives. Canadians were being killed and wounded at a rate faster than volunteers could be found to replace them. The solution was the Military Service Act, which proposed to conscript able-bodied unmarried men with exemptions only for those employed in jobs important to the war effort.

From its inception, conscription carried the seeds of structural division. For example, men in Quebec married, on average, at a much earlier age than men in the West. This required the initial burden of conscription to fall more heavily outside Quebec. Westerners who were already bearing the brunt of casualties seethed with resentment. Some wrote to their MPs demanding that French-Canadian "slackers" be hanged. When police wrongly arrested one Quebecer who had accidentally left his exemption papers at home, violence erupted. A Toronto regiment was sent to restore order in Quebec, rioters shot at the troops and they responded with machine guns and a cavalry charge with sabres drawn. The English press played up Quebec's anger as treasonous sympathy for the Kaiser. It proved the source of a bitter resentment towards Quebec that fulminates to this day in western Canada.

Labour, on the other hand, found memberships dwindling as men left for the front. On Vancouver Island, the Big Strike in the coal mines abruptly ended in 1914 when war was declared and striking miners enlisted beside scabs in the regiments sent to break the union. But when a 1917 strike at the Trail smelter failed, military police hunted and shot its pacifist organizer. Canada's first general strike followed when angry workers shut down Vancouver. In Calgary, 60 percent of trade unionists had enlisted and unions were forced to surrender their charters. The drive by owners to exploit an opportunity to disenfranchise union leaders was seen, perhaps rightly, as a conspiracy to use conscription to crush organized labour. It proved the source of deep suspicions toward business motives that still complicate industrial relations in BC.

Conscription also fostered divisions between rural and urban communities. The government had promised not to conscript farmers' sons. The farmers voted enthusiastically for conscription—since it didn't apply to them. When Ottawa broke its promise, 5,000 angry farmers marched on Parliament and anti-conscription disturbances erupted in Toronto and Calgary—although they appear to have been studiously downplayed by the same newspapers that thundered so righteously about Quebec's "slackers." This resentment and mistrust of urban Canada by its rural communities is still evident today in the rhetoric resource-based communities direct toward the "big city fools" perceived as the source of all their problems. If the Great War brought explosive industrial growth to Ontario and Quebec, it initially meant the opposite in the western provinces, where food and primary resource extraction took precedence over all else. This became the source of yet another bitter contemporary resentment by the west of a central Canada that is accused of commanding more than its fair share of the nation's industrial wealth.

What then is the truth of Walhachin? The fatal collision between romantic folly and a hard and unforgiving landscape? The fact that soil surveys later found not one hectare suitable for fruit trees? But then, I've come to understand that the truth is seldom to be found in the facts. It is almost always discovered in the heart. And beating at the heart of Walhachin's fleeting history is the emergent myth of Gordon Flowerdew and the relentless scythe of the Great War that swept him and his settlement away.

Gordon Flowerdew was born at Billingford, Norfolk, into a family that had been stewards to the Duke of Norfolk. He was educated at Framlingham College in Sussex. Like so many young Englishmen of his generation, he went to the frontier in search of his fortune. In 1911, riding his horse Dixie, the handsome young storekeeper won most of the trophies at the races held in Walhachin to honor King George V's coronation. In 1912 and 1913 he was conspicuous in the difficult Victoria Cross Race—an event in which mounted riders were expected to rescue a wounded comrade under simulated fire from artillery—that concluded summer cavalry training at the Vernon militia camp. But in 1914 he found himself in a war in which industrial technology had made cavalry obsolete, just as it would the tranquil farm economies of towns like Walhachin. In France, chivalrous young men came dreaming of glory and ended living like sewer rats. The howitzer, the tank, the truck and the aircraft were the emerging weapons of mobility. Time and again the cavalry would be brought to the front, then sent back to the rear as useless. Indeed, it was increasingly clear that the idea of cavalry had proved a gigantic drain on the war effort, tying up highly trained soldiers while its commanders fought a nasty political war to defy the Army Council's attempts to disband mounted units.

From the margins, Gordon Flowerdew fretted and fussed with his fellow officers at being left out of any chance for honour in a war fought by infantry and artillery. Many transferred out of horse to the Royal Flying Corps or the new tank units. Flowerdew was sent to the headquarters staff of the Canadian Cavalry Brigade. He was later rewarded with command of a squadron of Lord Strathcona's Horse, one of the five cavalry brigades attached to Sir Henry Gough's Fifth Army. It was the Walhachin rider's luck to be on leave in England when General Gough ordered the cavalry regiments to dismount and form reserve infantry behind the front lines.

At 5 a.m. on March 21, 1918, with a dense fog masking the Very flares and a devastating artillery barrage rolling before it, the

Kaiser launched Germany's last great offensive directly at Gough's positions. Almost overnight the dismounted cavalry became the front lines. Within days the Fifth Army ceased to exist as a fighting force. In desperate rearguard actions, both mounted and on foot, Lord Strathcona's Horse covered what was officially known as The Great Retreat but might more accurately have been called The Great Rout. The Germans had overrun 600 guns, taken 30,000 prisoners and were threatening the vital railway junction at Amiens which would open the way to Paris. Field Marshal Sir Douglas Haig ordered his troops "to die where they stood."

On the morning of March 30, the Canadian Cavalry Brigade was ordered to a frantic counterattack to seize a wood known as the Bois de Moreuil which the Germans were attempting to occupy. It was a crucial position because from it the enemy would have a commanding view of Amiens and the railway to Paris. For once the army would need their horses to get them there in time. C Squadron of Lord Strathcona's Horse was ordered to deny the heights to the enemy. Private Frank Richmond had enlisted with the regiment at age fifteen. As they cantered forward, what remained lodged in his memory until he died in Victoria eighty-eight years later was how closely the countryside resembled that open western parkland of British Columbia.

When Flowerdew led Richmond and the rest of his four troops around the wood at a gallop, he encountered two lines of enemy soldiers advancing with machine guns on both flanks and in the centre. He ordered one troop to dismount and make a diversionary flanking movement. The other three troops he led in a frontal cavalry charge against the machine guns. His shining moment had arrived. This was the Victoria Cross Race for which he had trained so diligently while a storekeeper in peaceful Walhachin. In the face of a withering fire from rifles, machine guns, trench mortars and howitzers, Flowerdew led Lord Strathcona's Horse to the attack at the full gallop with glittering sabres raised and courage that observers compared to the Charge of the Light Brigade.

He had lost 70 percent of his men when Richmond and the other survivors reached the German lines, passed through, sabring the gunners there, then wheeling to attack again. Although seriously wounded in both thighs, Flowerdew then ordered the remains of his squadron to dismount. They attacked again on foot, fighting hand to hand. The Germans fell back. Amiens and the way to Paris were denied. Did the gallant action of the Walhachin rider change the outcome of a war that would reconstruct the world? The old men who told me about it certainly believed that he and his little town had been placed at the fulcrum of destiny.

Gordon Flowerdew had led the last real charge of cavalry and in that sense his accomplishment marked the dying of one world and the birth of another. The horse had given way to the internal combustion engine. Women had entered the work force and won the vote and would transform the role of government. Across Canada, the equilibrium shifted from rural agriculture to the assembly line. Already the great migration from country villages to industrial cities had begun. By century's end, 80 percent of us would live in a few dozen urban sprawls and the empty countryside would be filled with place names that were once thriving communities.

On March 31, 1918, Frank Richmond was shot from his own horse by a strafing warplane. The same day, his gallant young commander from Walhachin died of his wounds, never to know that his last ride had won him the Victoria Cross, the one final trophy he had despaired of ever winning. In a sense, the village of Walhachin died with Gordon Flowerdew and the rest of those elite western riders who fell before the machine guns at Bois de Moreuil.

Many men failed to return to Walhachin after the war, widows left, the few soldiers who did come back—among them Gordon Flowerdew's brother—seemed changed in strange ways, unable to cope or care. The flume fell into disrepair. Unseasonable rains washed it out. The fruit trees, deprived of water, withered and died. Within five years of the armistice, the last settlers were gone. Today the elegant Walhachin Hotel, the polo matches, the card games in the lounge, the languid romances of that last serene and golden summer, all are ghosts that reside only in memory. The orchards are reclaimed by grass. Cattle graze where the night wind once whispered over a sea of leaves. Eden is again the dominion of dust devils—the same whirlwinds that danced for the Chief of the Dead before any European foot trod this country, the same wind eddies that began their fateful dance for Gordon Flowerdew and the dreaming town of Walhachin on August 4, 1914.

Bush Telegraph

"Guy wants to go down Malacca Passage some place. Lookin' for a ride." I put out a whisper on the bush telegraph and waited. Overhead, the million or so Prince Rupert seagulls roosting on the roof hunkered down over my room to wait with me. They rode in on a mean north coast sleet, harried off Hecate Strait by the dirty weather. Even that couldn't damp the squawks and squabbles upstairs—or the hoots and yowls of the midnight rowdies straggling out of the Belmont beer parlour just up the block. Gulls down on the roof in freezing rain, the incoherent yelps of Saturday night drunks skittering on a lacquer of clear ice—no matter what your language, those signs say nobody is going anywhere for a few days. Well, these things do take their time. You can't sit by the phone waiting while the world goes round. So I took my four-by-four and went walkabout.

Strange how a whisper in the right ear can reach out across the ether and tap you on the shoulder when nobody knows where you are, not even your surly editor back there in the Metro Nerve Centre. The message found me 300 kilometres up country, in a log cabin not far from where the Babine empties into the Kispiox. My hostess laughed out loud: "*Everybody* up here knows where you are!"

A kid at some Prince Rupert number where nobody lived told me: "Be on the Fairview float at first light day after tomorrow. If she's blowing hard, go home."

The morning was still and brittle as glass when I clambered aboard the *Equinox*, bound for Oona River in the care of the man who built her almost twenty-five years ago. Four fingers of snow dressed the hatch covers and the flakes hissed like something scalding hot where they settled into the black mirror of the sea's face. "Yeah, I built her. First we logged the timber, then we let it sit for a year, then we built her. I didn't want no steel boat, you know. I wanted a wooden boat."

Mike Lemon went to Oona River to build his boat because that's where Fred Letts built boats—famous boats for the halibut grounds and the Skeena salmon runs. For half a century he built them by hand, scouting the ideal crooks of timber, seasoning it, shaping it into boats like the *Diamontina, Oona Maid, El Nino, Wildcat* and *Blaze.*

Later, sitting in Fred's spartan kitchen, the old man shows me his pride and joy: the *Petrel C*, named for the channel where he found a log on the beach and saw her keel in it, right there, in one glance. A blunt, square-knuckled finger bangs down. "That's with 10,500 pounds of halibut down her hold. Look at the freeboard on her yet! My boy Bobby wants me to build one more." He snickers. "I'll have to move damn quick. I might not be around that much longer."

Fred was born in these north coast outports before the First World War. He jokes about his age because he's not ready to leave quite yet. In 1986 he went to St. Paul's Hospital and they stitched a pig's valve into his heart. Now, although seven decades of Skeena weather crinkle his face like a net mesh, his iron-grey brushcut and agile walk are those of a man twenty years younger.

But I'm getting ahead of myself. This story is still back in the wheelhouse of the *Equinox*, jammed in beside the six-and-a-half foot bulk of the former basketball star who is her skipper. Mike Lemon is bashful about his career as a national class Senior A player in Vancouver.

"Yeah, I played in a few national championships. They'd find you a job in Vancouver. Job didn't amount to a hill of beans but we were young and loved to play.

"You know, we had nothing in Prince Rupert when I grew up. We had the old YMCA gym. But we had so much fun. We used to go on boat trips to Kitkatla, Hartley Bay, Ketchikan. Some of those villages, they had nothing, but they had a basketball court. We'd get billeted out in the community. On the court we weren't treated so special, but after the game we were royalty.

"Those Tsimshian teams—they loved to run and gun. Fast break. Shoot. Fast break. So we'd slow the ball down. Play control. They'd get so frustrated. They used to switch the lights off when we had the ball. 'Might as well sleep,' they'd say."

What beached the star athlete in this last of the outports? This remote settlement where tides still decide who comes and who goes and there are no phones, no stores, no doctors, no power. Fred Letts, it turns out, had more than a boat slip and an eye for spotting the perfect keel hidden in a yellow cedar log. He had a beautiful teenage daughter. Mike stayed, he says, because it was in his mind to marry Jan. Oh, there was power in Oona River, all right. The power of the heart.

Not so long ago, if you wanted to pay a visit here, you had exactly two hours on either side of high tide. If you missed the window, you stood off for eight hours and waited for the next one. Or chugged back to Prince Rupert, 40 kilometres to the north. "So what?" grins Fred, village patriarch, resident philosopher and builder of superlative fishboats. "Waiting two hours for a tide or waiting two hours for a plane—what's the difference, exactly?"

Times have changed a bit since the federal government put in a small breakwater and dredged a channel a few years ago, but they haven't changed much. The window for getting in and out of the moorage basin is a few hours wider, but even the weekly mail plane from Prince Rupert comes in on floats only if the tide is right.

Hauling my sleeping bag and winter gear up from the float while Mike rustles up somebody who'll put me up for a few days, I lean on the rail to take stock of my surroundings. Oona River is a slow-moving tidal estuary that fans out through muskeg and peat-bogs on the low-lying southeast shore of Porcher Island, one of the biggest islands with the smallest populations among the thousands scattered down the British Columbia coast. Behind me is the south end of the Chismore Range, a strange-looking mountain that Oona River settlers call either Mummy Mountain (from the sea it resembles a pregnant woman) or Mt. Kewpie Doll—for the fat-bottomed plastic dolls you used to win at the PNE. From up there, I'm told, you can see the distant Queen Charlotte Islands on a clear day. Dispersed for several miles up the muddy banks, their home-steads stretching back into the scrub timber, are the dozen or so houses of the settlement. They are set well back from the water, the houses of people still aware of the power of the elements and prepared to sacrifice the aesthetics of the foreshore for a good windbreak and a south-facing slope that catches the winter sun. The skyline is uncluttered by telephone poles or power lines. That's because there's no outside source of electricity and no private tele-phones here, although that will change soon. For now, a single BC Tel radio telephone perches in its stunningly incongruous blue and white box outside the schoolhouse. Householders, however, yak together on what they call the Mickey Mouse—the low cost citizen's band radios they've strung together on the same frequency. It's a big improvement over the hand-cranked war surplus army field phones they used before. You notice right away that there are no licence plates on Oona River vehicles—why would there be?

There's no road, except for the muddy track that runs a mile back up the river from the government wharf.

Ralph Letts gives me the grand tour in his truck, a hybrid monstrosity salvaged after some handyman welder abandoned it in the Alaska State Ferry lineup. A pal barged it over. "What the heck," says Ralph. "It runs in low gear. That's all you need around here." When it dies—and right now it sounds awfully sick—it will join rusty ranks of equipment, dead cars, pipe lengths, empty drums, piles of junk metal, beached boats and scrap lumber along the road.

"People come here and they think it looks kind of like the dump," my guide muses. Ralph is reading my mind. "You have to understand that all this stuff in the blackberry brambles is kinda like our hardware store. At Oona River you can't just run down to the machine shop and order a part, so you never throw anything away. You never know when you might need a two-inch piece of this or that." He gestures at the scrap pile behind the settlement's tiny sawmill. "You need something, you just come down here and help yourself."

I can't imagine what I might need along those lines right now, but I appreciate the offer. The cooperative spirit of this community reaches all the way back to its roots. It was settled just after the turn of the century by bachelor fishermen who liked the sheltered anchorage. More important, in their view, was the fact that Oona River is directly opposite the three vast channels of the Skeena and its once-immense salmon runs.

At Fred's place, we sat around the kitchen table and I was reminded again that the best stories emerge not from the glitzy billion-dollar yammering of the boob tube, but from the yellow glow of a kerosene lamp and the yeasty scent of fresh sourdough rising by the woodstove. You come across good stories in places like this, where people still know how to entertain each other; where they cherish the mother arts of storytelling, conversation and song.

And bake their own bread.

It was a cold winter's night with more than a taste of snow in the wind. The white bones of moonrise cast long shadows down Bareside Mountain. Fred Letts pushed aside jars of his special huckleberry preserve and wild raspberry jam, cleared a space for his elbows among the dinner plates and leaned forward to share a mystery from a youth spent more than half a century ago.

"He was there when we came in 1924. He'd been here earlier and then disappeared for a while, but he came back. He lived about four miles up the coast from here towards Spiller River—a heavy-set French-Canadian fellow. He had a real French accent. Oh, he was a French-Canadian all right, no doubt about that. He must have been pretty near six feet and although he was 55 or 60, he could still travel around pretty good.

"The one thing I remember, he used to chew gum—spruce gum. He'd chew that stuff right out of the tree. After a while it gets harder than hell, then you have to start another piece.

"This guy was a real loner. He lived by himself in a little log cabin, about eighteen by twenty feet, and he got around in a 16-foot clinker-built rowboat. He'd row over to Oona River every once in a while to buy some tobacco. Old Man Anderson was one of them Swedish bachelors that settled here and would sometimes row over there and visit him for a couple of days."

The lone wolf's name was Joe Levine and he had been, Old Man Anderson swore to the boys of Oona River, a real hard case in Soapy Smith's gang. Jefferson Randolph Smith was the outlaw king of Skagway at the height of the Klondike Stampede. Card sharp, confidence man, Smith had preyed on miners in gold camps from Colorado to the foot of the Chilkoot Pass. In Skagway, he hit the big time. He bought the marshal and his gang held the town in an iron grip. Sam Steele, commander of the Mounties on the Canadian side, passed through when Soapy reigned supreme: "Little better than hell on earth," Steele wrote of Skagway. "It seemed as if the scum of the earth had hastened here...There was no law whatsoever; might was right, the dead shot only was immune to danger..." Robbery and murder were daily occurrences...Shots were exchanged on the streets in broad daylight...At night the crash of bands, shouts of 'Murder!,' cries for help...and occasionally some poor fellow was found lying lifeless on his sled where he had sat down to rest, the powder marks on his back and his pockets inside out." Whether greenhorn or old sweat, nobody passing through Skagway was safe from Soapy Smith's gang of bunco artists, stick-up men, four-flushers, pimps, whores, goons and bushwhackers.

In July of 1898, Smith was killed by Frank Reid in a gunfight on the docks and his gang was being hunted down by a lynch mob. Joe Levine had good reason to go to ground in the quietest place he could find. A lot of tough folks had scores to settle.

"I guess all these islands have seen a lot of people coming and going. It was pretty easy to disappear around here. A lot of the early settlers were Finns and Swedes avoiding conscription back home. During the big war, the first one, a lot of them hid out up the river.

"It didn't cost much to live here. There were deer all over the place. Fish. You could shoot your grub half the time, catch it the other half. This guy Levine, he had a big garden, huge rhubarb plants. He'd go jigging and handlining out of that rowboat for the larder, but he got his cash hunting seals—that man was a crack shot.

"One day in 1935 or 1936, Old Man Anderson came in with the strangest story. He'd rowed over to Levine's cabin. Everything was neat as a pin. The table was set for dinner, the pot cooking on the stove, boat tied up at the beach—but Levine was gone. Vanished. Not a trace. We never saw him again."

Did some Yukon sourdough come to square a debt?

"Who can say? Guys like Levine were alone a lot. If he fell in and drowned nobody would know—the crabs wouldn't tell. Funny thing is, what I remember best is Joe Levine's doughnuts. He'd fry 'em up in seal fat. I was surprised how good they were—they didn't taste fishy at all."

That night, fattened up on a dinner of boiled winter spring, potatoes and berry preserves, I dossed down in the middle room of Fred Letts' homestead. The trees sighed and a corner of loose tarpaulin slapped. Ghosts stirred in the wind. George Letts had come west with his wife Martha in 1907 to escape desolate Yorkshire slag heaps and a life in the dark, cutting coal. What he found was this luminous world of sea and sky, wild mountains, the bounty of the great river. Four generations later, his descendants are still here, moved by the same rhythms of season and weather, sharing the same cloak of wisdom, humility and generosity that make Oona River as big, in its own way, as any place you'll ever want to go.

Grey Shepherd of the Log Booms

Untended in the clutter of my basement workbench lies a forearm's length of twisted steel, a pocked reminder of the impermanence of cenotaphs and other material things. The edges are ragged and razor sharp. Rust flakes off in pieces as big as your palm. Only across one broken end does it gleam with the bright memory of its uncorroded core. I pried it loose from a decaying bulkhead one bitter day—my personal Remembrance Day—as squalls of freezing rain and snow walked the iron-grey sea. These storms lash Vancouver Island's exposed shoreline every November. They scour Kuhushan Point and boil over the bar at Black Creek and send the white manes of their horses tossing all the way from Lasqueti to Mittlenatch. I persevered. Chest-deep in icy, foam-streaked swells, hands numb and knuckles bloodied by barnacles and the rusted stumps of rivets, I braced against a flood tide rising fast on a Strait of Georgia sou'easter and pulled and pulled until the piece came loose. The discomfort earned me all that remains of the *K444* except for rusted fragments strewn across a weedy seabed.

I'm one to dismiss sentimental attachments to things. Our

culture is obsessed with its icons of the past, a vast reliquary of pop culture knick-knacks. Cupboards pile up with plastic cups from the last *Star Wars* movie, cartoon characters from Disneyland, collector plates that deify dead princesses. We commemorate visits by actors, popes and rock groups, all the while collecting banal junk, the inappropriate, the entirely false, while our real history decays around us. Still, I trace descent from able seamen who served British dreadnoughts like HMS *Revenge* and HMS *Lion* under Victoria and Edward. I accept my own nostalgia for a wreck that charts the lost latitudes of childhood.

The *K444* is a kind of *Flying Dutchman* that sails forever on boyhood dreams. She was always a magnet for any kids who could persuade parents driving north from Qualicum on the old Island Highway to stop for a picnic on the windy spit where it hooks out into Oyster Bay. On a sunny day, the prospect is timeless and spectacular. Lined with driftwood like the bleached bones of primeval creatures, the spit bristles with life. Pebble beaches curve away to north and south below rustling seagrass and nodding goldenrod. Behind, the gentle coastal plain gives way to the alpine meadows of Forbidden Plateau and the soaring peaks of Mount Arrowsmith and the Golden Hinde. Across the tide-ripped channel lies the awesome wilderness of the Barkshack River country, a vast arc of devil's club and volcanic outcrops from Tantalus to Hat Mountain. Texada Island humps out of the straits, tilting north to Blubber Bay, where Elijah Fader cut up whales in 1890. Farther north brood Cortes Island and the Redonda group, with Whaletown, Refuge Cove and Desolation Sound.

Like many a child, I knew the *K444* only as a hulk, stranded at the edge of a mud flat. With them, I share the spinning of stories to explain her presence. Depending on the year and current reading matter, she might have been beached by mutinous pirates on a smuggling run from Singapore, lost in a hurricane, or rammed at full bore into the shoals by a drunken crew—as was the brave little Hudson's Bay Company paddle steamer *Beaver* in 1888. In truth, during post-war decommissioning, the navy (we were sure we'd never need one again) simply scuttled *K444* to make a breakwater for the Oyster Bay booming grounds, collecting point for timber coming out of the Iron River with the first of the truck loggers.

Forty years ago, the *K444* still looked like a ship. Her superstructure was gone and the salvage yard had stripped off her running gear, but to north-bound travellers, the raked sides and their tall white numbers were a landmark. They loomed out of rain and spindrift that gusted off the bay and up, over the Island Highway, into the second growth. That image transfixed the childhood me. A shipwreck. A real shipwreck. One you could walk to at low tide, crunching through clamshells, sliding naked toes through eel grass, stirring the baby flounders and braving the stench that oozed and belched from the bottom with every step. What girl or boy could resist such a temptation? Mysterious, strange, the *K444* was a lodestar for young imaginations fuelled by Hollywood matinees and pulpy pre-war adventure novels borrowed for summer reading from bookmobiles and circulating libraries at Port Alberni, Courtenay and Campbell River. That hulk was encrusted with tales: wolf eels lurking in the wreck, huge and fierce, capable of biting off an unwary arm in a single crunch of the jaws; ghostly lights on the foredeck at dusk; the sounds of voices echoing through slimy companion ways; kids lost below decks in the rising tide—pale, drowned faces bobbing like jellyfish at submerged portholes if you could only swim down to see. Tide after tide, storm after storm, she seemed more real even as she vanished into herself, steel plates buckling and bulkheads rusting.

Marine authorities somewhere finally decided the *K444* was a navigational hazard and blew her to smithereens, dispersing generations of dreams with a few well-placed dynamite charges. Even at that, the keel clung to the spit for awhile. And it was to that last, pathetic remnant that I made my pilgrimage, descending from the high plains to claim a piece of my lost island youth. Only later did I come to understand that the rust-scabbed steel in my basement is more than a charm to conjure lost childhoods. It has become my own personal cenotaph, a talisman. Later still, searching for the truth, I played years of hide-and-seek with a ghost ship. She led me through Canada's national archives, marine records and correspondence with naval veterans. In the end, her tale was no less exciting than those we made up as kids, playing into our futures beside a disintegrating past.

K444's keel was laid at the Canadian Vickers shipyard in Montreal during the darkest days of World War II. The Axis was ascendant and the Battle of the Atlantic was in full fury. The Nazi wolf packs came to starve Britain, the precarious toehold of democracy in Europe, and U-boats harried the supply lines right into the St. Lawrence River. In the spring of 1942, the grey submarines were hunting as far upstream as Rimouski. Between May and October of that year, they sank twenty-three ships including three warships: HMCS *Raccoon*, HMCS *Charlottetown* and HMCS *Shawinigan*. On October 14, a U-boat torpedoed the Sydney to Port au Basques ferry, killing 137 people. It was a campaign prosecuted at a frightful cost. Before the war was over more than 3,000 Canadians would die in the battle to keep North Atlantic sea lanes open. The U-boats would sink 3,600 merchant ships and 175 warships during the war, but at a cost of 800 submarines—a casualty rate of 75 percent.

They launched the *K444*, my ship of the imagination, as HMCS *Matane*. She was one of the Royal Canadian Navy escorts that broke the German submarine offensive in the North Atlantic. Twenty-four of them would be sunk but Canadian ships would sink twenty-seven submarines and Canadian aircraft twenty-five more. Ships like the *K444* escorted merchant convoys that carried 180 million tons of supplies across to beleaguered Britain. With a top speed of 18 knots and a crew of 136, HMCS *Matane* was a River Class frigate armed with four-inch guns fore and aft, two Oerlikon anti-aircraft guns, standard depth-charge gear and the new Hedgehog that fired precise anti-submarine patterns. It's more than half a century since the lean, fighting frigate put on her warpaint, slipped past the Quebec fishing village that provided her name and into the St. Lawrence to accompany Convoy SQ71. From the smoking shoulder of Iceland, through the frigid waters of Greenland and down the ice-choked sea lanes to Bloody Foreland and the Arran Isles, she's assigned to the Western Approaches with Escort Group 9, bringing seven-knot convoys through the fog and sea smoke. And hunting U-boats with deadly efficiency.

I found her logbooks buried in a warehouse outside Ottawa. As far as I know, I'm the only one to open them since they were closed by the last watch—years before I was born. Sitting in the comfort of the National Archives reading room, looking out on the fall colours along the Ottawa River, I felt only the chill of tension and terror in laconic entries under wartime skippers—Lt.-Cmdr. Easton, Royal Navy Cmdr. Layard, Lt.-Cmdr. Lucas. Are they still alive? I wondered. Do those cold moments enter their night thoughts? These Canadian crews, mustered from farm boys and kids right out of high school, excelled at deadly battle tactics. Their most sinister skills emerge from the brief entries in logs that tell of strength-sapping cold and mind-crushing boredom; of men straining, watch after watch, to see into the darkness and spindrift; of the trembling of a greyhound forced to idle among lumbering merchant ships, listening for that frightening, inevitable ping of an ASDIC contact; the shocking orange blossom of a tanker burning against black sky, black sea.

Like the memories of kids who reinvented history to serve childhood games, the precise entries of real history now brim with mysteries. The entry for March 17, 1944 reports many submarine contacts but no action—just a small, abandoned boat drifting on the high seas. Later, there are reports of strange star shells, flares, gun flashes suggesting a sea battle beyond a horizon where there are no known ships. On March 26 she sights her first submarine. That night, HMCS *Matane* erupts from 6 to 16 knots; sprints to station then halts suddenly to listen; shadows her sister to mask the number of warships in the escort; races ahead of the convoy and stops all engines; drifts back to meet it; lays silent traps on the dark side of moonrise to bushwhack an over-ambitious U-boat captain. The ship of childhood games turns out to be a ruthless killer.

On April 17, 1944, she ambushes a submarine at 8:29 p.m. At 8:33 her log reports "heard underwater explosion." On June 29 at 2:46 p.m. another attack and another "tremendous explosion" below, with debris, clothing and US blueprints rising to the surface. She takes part in engagements that kill the *U-845* and the *U-448*. Yet there is no crow of triumph in these records, only weary relief that the dead are other young men like those of HMCS *Matane*'s crew—not *our* dead.

Later, the ebbing tide of Nazi fortunes carries *K444* across the Atlantic. Caught without air cover and perilously close to Fortress Europe, she is attacked by the Luftwaffe in the Bay of Biscay on July 20, 1944. Her hull breached by a glider bomb, gun decks out of action and heavily damaged, her engine room flooded, with two officers and nine seamen wounded, the *K444* refuses to sink. At 5 a.m. the following morning, the first of the wounded dies. Like so many of Canada's best sailors, Able Seaman John Cole came from a whistlestop in the dusty prairie. In Stranraer, Saskatchewan, he leaves his Elinor a widow. Sister ship HMCS *Meon* takes HMCS *Matane* in tow to Britain's shipyards for repair. The damage is so extensive that it takes eight and a half months to get her back into action. She returns briefly to active duty on the Murmansk run, escorts Russian supply convoys around Norway's stormy North Cape to the White Sea and the port of Archangel.

And then, one May morning in 1945, wireless operator Ron

Bailey picks up an emergency signal from a clandestine naval station in Norway to the Admiralty in Whitehall: "Have observed convoy of 15 U-boats and five surface vessels outbound Narvik. Is this in order?" The next message orders HMCS *Matane*, in command of the frigates *Nene, Loch Alvie, Monnow* and *St. Pierre*, to intercept the German fleet. "We encountered the U-boat fleet surreptitiously making way down the Norwegian coast towards the Baltic led by the most beautiful vessel I have ever seen," he writes later. "Once Hitler's personal yacht, *The Grille* now served as the mother ship of the U-boat flotilla. She had the classic lines of a clipper ship."

Here, off the coast of Norway, the *K444* earns her formal place in the great tide of history. Lieutenant J.J. Coates' boarding party to Hitler's yacht accepts Herr Kapitan Lieutenant Suhren's formal surrender of the remnants of the Third Reich's Arctic and Barrents Sea fleet. "He was then ordered to inform his U-boats that if they attempted to scuttle or submerge, the boat would be destroyed and no mercy would be shown to the crew." On May 8, just off the Faeroe Islands on the long, slow journey to Scotland with the German fleet, the radio officer takes a message. It's from Winston Churchill himself:

> On this great day I send to every officer and rating in the Royal Canadian Navy and the reserves my personal congratulations. The part you have played to help accomplish this result has been valiant. Now we can look forward to an end to the tedious work in the convoys, to the hazardous hunting in the support groups, to the fights by night in the destroyer flotillas. Germany is beaten again. In this moment of victory I know you will be saddened when you think of those who have given their lives that we all might live once more in peace.

The ship's company was mustered, "Up Spirits" was piped and the captain issued a double tot of grog. He proposed a toast to victory. His German counterpart responded: "To our unvanquished foe—the sea." The fate of his ship remains a mystery, too. Officially, she was broken up by the British in 1951.

But Ron Bailey, writing about his experiences fifty years later, says that in 1955 while riding the ferry to North Vancouver, he saw the silhouette of "an exquisite yacht" at anchor in Coal Harbour. "I could not be mistaken. Those elegant lines from half a world and decade removed were indelibly etched in my memory. *The Grille* was still afloat. She enjoyed a kinder fate than *Matane*..."

With Escort Group 16, the *Matane* had accompanied the last troop convoy of the war to Gibraltar and the last one back to Britain, then sailed for Panama Canal and passage north to the dockyard at Esquimalt. Born in Quebec, not far from the birthplace of the nation she defended, her active service ended in BC in 1946, only three years after she slid down the skidway and into the St. Lawrence. The last entry in her log reads: 21:00 Rounds Correct. 22:30 Pipe Down. Sold as scrap to Capital Iron and Metals Ltd. in Victoria, her superstructure was stripped, her engines removed and the hulk was sold to the Iron River Timber Company. She took up final station as *K444* on the beach at Oyster Bay the next year.

On Remembrance Day, I'll sometimes forego the official cenotaph ceremonies—not out of disrespect. Instead I revisit childhood memories and drink a toast to my rusted steel, to the *K444* and her place in the heart. Grey shepherd of the log booms, escort of children's dreams. Her last patrol was the best one.

Annie's Variety Store

Red hair flying, Annie Tynjala comes bouncing across the road in her blue track suit. Lively as a lacrosse ball, she has the rounded contours of a woman who's just celebrated her seventy-ninth birthday but fizzes with the energy of someone thirty years younger. At the door, framed by clapboard and casements, Annie pauses to flash a single brilliant smile, then disappears like the White Rabbit down his hole. Following her inside, I discover a glass-fronted storekeeper's case from a gentler age and the engaging clutter of everything I imagined the perfect store would contain when I was a kid, from Buzz bomb bumpers to racy magazines. The store is open, but it isn't. Or maybe it is. On the other hand, Annie's just about to close—a neighbour's coming to make sure she gets her regular check-up at the clinic. This whole experience is taking on an Alice In Wonderland quality.

To find Annie and her Variety Store, you catch the ferry from Port McNeill on the north end of Vancouver Island to Sointula. Enjoy Captain Hook's bullhorn on the way. He uses it to bellow from the bridge at rambunctious school kids. On this day, a clammy north island fog sinks to the water. Beyond the ship's rail the hulls of drum seiners ghost past, watching with electronic eyes what I can't see. The mist-draped treeline of Malcolm Island emerges from the fog bank like the fabled isle of Avalon. Tiers of brightly coloured houses curve along a shelving beach, the branches of fruit trees spread above carefully groomed gardens still bright with flowers. It was supposed to be Avalon once, the blessed island in the western sea, a haven for those fleeing a world that persecuted their religious beliefs and their politics.

Matti Kurikka dreamed of founding a heartland for Finnish culture and traditions far from the oppressions of the Russian czars and the capitalist robber barons who ruthlessly exploited immigrant labour in the mines and logging camps. "In this colony a high, cultural life of freedom would be built, away from priests who have defiled the high morals of Christianity, away from churches that destroy peace, away from all the evils of the outside world," he wrote.

The Finnish community at Nanaimo, unhappy at the manipulations of Premier James Dunsmuir, who moved his mining towns at whim—and the workers always found they had to buy their new building lots from his company—asked Kurikka to make his dream a reality. He chose Malcolm Island, with more than 11,000 hectares of cedar, spruce and hemlock forests, about 200 kilometres northwest of Vancouver. The island straddles the entrance to Johnstone Straits, a critical choke point for the teeming runs of salmon. The fish sweep in from the North Pacific, first for the Nimpkish River and its tributaries, then the mainland inlets, and finally down into the Campbell, Oyster, Qualicum and Fraser watersheds which drain into the Gulf of Georgia. Here the land rises in gentle contours from curving sweeps of beach—ideal for later clearing and farming. With forests and fisheries at hand and rich potential for agriculture, it seemed ideal for building the Finnish utopia.

In this new settlement, women were to have full equality with men in all community matters of politics, commerce and law, a daring vision indeed almost a generation before the first Canadian women got the vote. Children were to be a community responsibility and the colony would bear their costs of food, clothing and schooling. But from the first day, the crude realities of the frontier intruded into Kurikka's vision of a perfect, socially engineered settlement where the devout might commune with God without the interventions of church, or priest, or the masters of Mammon.

A preliminary settlement expedition arrived in late fall of 1901 and stumbled into the ruins of an earlier utopia abandoned to the blackberry brambles and salmonberries. An earlier religious sect had already foundered in its dream of a New Jerusalem. To make things worse, the new arrivals lost their provisions in a heavy sea and spent the first night in Avalon huddled under a tree, bitterly cold, warming their littlest children against their bare chests. That winter, Heikki Kilpelainen and Viktor Saarikoski were dispatched to build a log cabin to house the colonists who would follow when shelter was available. By March of 1902 there were fourteen men and Mrs. Anna Wilander, who had come with her new husband by sailing ship. A year later, the settlement felt sufficiently established to charter the steamship *Capilano* and bring a large party from Nanaimo. The vast stump of a cedar felled to provide timber for the settlers' housing served as a stage for celebrations of the founding of their dream. Their beautiful, isolated island would be one of prosperous dairy farms, socialist collectives free from wage-slavery, and stunning triumphs of the mind that would advance Finnish culture. At first, they planned to name the place Koti, or "Home," but that was rejected as too utilitarian for utopia. Sointula, "Harmony" in the language of Suomi, was chosen instead.

But timber prices fell. So did the price of salmon—Rivers Inlet fish were bringing only seven cents apiece, regardless of size. And the provincial government had granted a cannery at Alert Bay a monopoly on the Nimpkish River run. In the winter of 1903, mired in debt and poverty, with internal divisions beginning to rend the fabric of Harmony, the colonist met in their three-storey communal shelter on January 29 to discuss the future. While the adults were meeting on the top floor and the children slept below, a fire started in the ground floor bakery. Two women, a man and eight of the settlers' children burned to death.

Yet it wasn't despair and grief that tore early Sointula apart, it was love. Or at least, the idea of love—Free Love, the new socialist idea that marriage was an institution used by a patriarchal state to enslave women. Matti Kurikka embraced the idea, in theory only, and wrote about it in his newspaper *Aika*, to the consternation of the wider world. The idea of a remote community of socialist foreigners founded on principles of free love created an uproar in starchy Victoria and the conservative citizens of Sointula were horrified at their growing reputation. After a special meeting of the colonists, Kurikka resigned and left the island, accompanied by half the community. He never returned. On May 27, 1905, at a meeting attended by thirty-six settlers, the utopian colony was finally dissolved as a formal entity.

Yet many stayed and set about building their lives as individuals. Men went away to work in logging camps and canneries, the farms flourished under the diligent hands of their wives, a school was built, the library was maintained and its collection of Finnish books grew to 2,000 volumes. In the end, the socialist dreams were lodged not in free love, but in the community's thriving co-operative store.

But all that's in the past, most of it beneath the blinding green turf of the immaculate little cemetery with its fishboat gear and carved salmon on the tombstones. Today we're trying to find Annie Tynjala, The Millionaire.

At the ferry dock, climb sharply to the white clapboard of the oldest continuously operating co-op store in British Columbia, perhaps Canada. Behind it is the Ode To Joy Bar and Lounge, just up the road from the Inn Convenience Store. Turn left and head past the Malcolm Island Inn. Great chowder, decor by Flotsam and

Jetsam: rusty horseshoes, an old grizzly bear skull grinning like a teenage pool shark, carrying yokes for water pails, fastball trophies. A stuffed marlin? Keep going, past the Finnish Organization Hall, the fire station, the justly famed library—still historic Sointula's prized possession. The road narrows, elbows through a weathered jumble of net lofts, boat sheds, repair shops, tangles of blackberry vines.

Arrive at KD's Kitchen, alias the Rough Bay Ritz, and you've gone too far. Here the chairs are six different colours, seven different species, and you can rest your elbows on pink plastic table covers and jot down fragments from the seinermen's chat: "Take a full moon and get the peak of the run." "There's too much gear in the water, but I've got to fish once more." The laconic chat is a reminder that despite the early setbacks, this is where the technology of the drum seiner was invented and transformed the west coast fishery.

After a coffee strong enough to strip barnacles from a double-ender's weedy bottom, I head back to Annie's, still closed until she streaks across to open up for me. Is it true she's a millionaire? That she owns a Stradivarius violin worth at least that? A Stradivarius played in concert at Carnegie Hall in New York in 1936? Eyes twinkle at the impertinence.

"It had the papers, 1721. I used to have it until 1987."

Did she sell it? For how much? Is that how she became a millionaire?

"I don't sell it. I give it—to the Music Academy at Valkeakoski, Finland. I put it in my father's name."

Jooseppi Wessman, child prodigy, 1887–1918, who had it from his father, Tuomas Tapaninpoika Wessman, 1859–1904. Her father left his treasured concert instrument to her brothers. One by one, they died. It passed down the bloodlines, travelling the far branches of a family tree until it arrived with Annie, the youngest, on the beach in Sointula.

"My mother wouldn't sell it. My brothers wouldn't sell it. It was not for me to sell it. It should be making music. The director, he tapped it on the back. He said: 'Oh, that will make beautiful music!' So I gave it to them—to make music."

She could have been a millionaire. So who needs to be a millionaire when all your neighbours love you and you run Annie Tynjala's Variety Store in a town named Harmony?

Battle Flag of the Orchards

Framed by age-darkened wood, easy to overlook in the side gallery of the Penticton museum where it's displayed, is one of Canada's most important historical treasures. It is a faded red silk ensign, simple and unadorned, the flag that flew when Canada was born on a windswept height called Vimy Ridge. Wait, you say, recalling your high school history, Canada was born in 1867. But a country is always more than its official documents and history. It is also something intangible that can only be forged in the collective imagination of its people.

Ask those who were there on April 9, 1917. They will tell you that while Canada may have been conceived in a political deal, it wasn't truly born until fifty years later in the snow at Vimy. And its midwife was not a polished statesman but a scandal-ridden real estate dealer from Victoria named Arthur Currie who happened also to be a military genius. "In those few minutes [at Vimy] I witnessed the birth of a nation," said Brigadier-General Alexander Ross. "I became a Canadian on Vimy Ridge," confessed George Pearkes, the British-born anglophile and Victoria Cross honouree who rose from lieutenant to command the army and later served both as MP and as British Columbia's Lieutenant-Governor. Vimy Ridge has now receded to the fraying edge of living memory. Soon it will pass entirely from the realm of human experience and become part of a world we can know only through archival reconstruction, like Waterloo and the Battle of Hastings. Even the youngest who fought in the Great War For Civilization are now in their mid-nineties. A few survive, waiting for the scythe of time to accomplish what shrapnel and machine gun could not.

Maurice Joslin of Penticton, 105, served with the 38th Ottawa Battalion. He is still plagued by the pain from a phantom hand. He was mutilated by machine gun bullets, his fingers amputated in the field with a pair of kitchen shears. "It's a mystery to me now," he says, holding up the maimed hand. "Well, it was almost suicide you know. You were almost certain you were going to get it. One of the battalions with us was a Highland regiment in kilts. A lot of them got killed. We went through their position. It was a strange feeling going past those dead kilties, lying there with their legs every which way."

Roy Henley of Sidney, ninety-five, among the 410 Canadians to win France's Croix de Guerre, was one of those Highlanders. Ask him about Vimy and Henley pauses. His wintry blue eyes lingering on the framed photo of ten handsome young men in kilts. Grinning for the photographer, burdened with packs and rifles, wire cutters, tin hats, gas masks, Lewis gun magazines, canvas aprons bulging with Mills bombs, they swing muscular legs through the tall grass of the open French countryside. "Most of those people were dead by two o'clock on the afternoon of that particular day," the old soldier says. "That's me, third from the right. We made Canada at Vimy," he says abruptly. "A lot of people, they get bored with us. They think: silly old buggers spend most of their time in the Legion. But we made this country for you." He recites the names of dead men. George Price, killed two minutes before the war ended. Private Croker. Private Daigle. Bob Scott, who went through the whole bloody thing without a scratch and then caught one in the last hundred days. Lance Corporal Jones. Men from the Windmills, Edmonton's 49th Battalion. "Oh, we were very tight with the Windmills. Tangle with a 42nd man while there's a 49th man around and you've got real trouble on your hands."

Then his eyes focus on something far away, something beyond this small room cluttered with memorabilia, his row of tarnished medals, cap badges, trench maps still marked "Secret, not to be taken beyond Brigade HQ," a sabre, the bugle which played the cavalry charge at Moreuil Wood where Gordon Flowerdew of Walhachin just west of Kamloops won his Victoria Cross. For a moment he is Private Henley again, a saucy fifteen-year-old soldier enduring the terrifying night of his country's birth. The night when the Canadian divisions advancing stealthily through a moon-silvered wasteland suddenly vanished into the dark hillside. The boy soldier found himself crammed into a dank tunnel. It stank of unwashed bodies, wet khaki, fear and—incongruously—the classrooms he had abandoned for this high adventure: around him, a tang of chalk dust hung in the air. Men were carving their initials into the soft white tunnel walls, leaving something of themselves, small talismans against extinguishment. Henley had lied about his age, enlisted with the cavalry in Montreal on a lark, then volunteered when the infantry called for replacements. He drew the Black Watch, still young enough to be shocked to discover the truth about fighting regiments. It was true: they wore nothing under their kilts.

The tunnels were part of an ingenious strategy devised by Arthur Currie both to mask his intentions from the overlooking enemy and to protect his men until the last possible second. This plain-spoken officer had assumed command of the four divisions of the Canadian Corps to prepare the attack planned by British General Sir Julian Byng. It was a risky, even desperate, venture. In

1917, Allied fortunes were at their nadir. The Bolshevik Revolution had toppled the Czar and Germany would no longer fight a two-front war. The French had exhausted themselves at Verdun and mutiny was brewing. The British had suffered a series of ghastly disasters on the Somme and at Gallipoli. Henley suddenly roots among his papers. "I have a Times of London here somewhere. It lists the second lieutenants killed on that day in July [at the Somme]. Their names fill six columns of the paper. The life expectancy of a second lieutenant was two weeks."

Over all this loomed the unassailable German-held heights. "It is impossible to over-estimate the strategic importance of Vimy Ridge," said one contemporary military report. "Its two spurs, flung out west and south-west in a series of heights which dominated the western plain, were regarded by military experts as the backbone of the whole German position in France." The ridge was also considered almost impregnable. Already 200,000 men had been spent on its slopes in four Allied attacks. The French, who had lost it in 1914, had been slaughtered attempting to recapture it in 1915. "The British tried to take it the following year. They have a big cemetery there to prove they couldn't do it," Henley says. "They even lost part of their own line in the German counterattack." This was the prospect facing that frightened teenaged boy so long ago: taking a redoubt that had already cost the equivalent of half the entire Canadian Army.

Like most of his men, Currie was a citizen soldier. He had never trained at a British staff college. His reputation was tainted by a pre-war scandal involving the misappropriation of $10,000 in a real estate speculation. But Byng had faith in Currie's other talents, particularly his common sense. This Canadian had learned from the Somme, where 21,000 British troops had been massacred in a single afternoon, leading the German high command to observe sarcastically that the Tommies might fight like lions but they were led by donkeys. So Currie modified the staff college theories that failed in practice. He employed Colonel Andrew McNaughton, a McGill chemistry professor, to calculate how to counter-target German batteries by measuring flash and sound.

For months, Canadian patrols were sent into the dangerously exposed No Man's Land of tangled wire, toxic shell holes and earth harrowed by shrapnel and artillery barrages. Their job was to probe the German lines and report. Henley was one who was sent out.

"Oh Christ, it was terrifying. Every bush, every shadow, snipers. I'm alive because I was scared and I stayed scared.

"We had one Native Indian sniper. He'd slip out at night and put up a stake next to the German lines. He painted our side of the stake with phosphorescent paint, then he sighted his rifle in on it. Whenever the glow of the paint darkened, he fired. That's what it was like for us, too. One of our snipers was killed by the sniper he killed. They shot each other."

This probing cost 1,400 Canadian casualties but later it saved tens of thousands of lives. Using the knowledge bought with the blood of his patrols, Currie constructed a vast replica of Vimy Ridge behind his own lines. And then every Canadian soldier, from private to general, drilled relentlessly in his specific task for the coming battle. Timing was crucial. Currie now knew that to avoid another massacre on the slopes he must change the way in which modern warfare would be waged. Instead of lifting his barrage for an orderly advance, he would roll it up the slope and his troops would follow in short rushes and try to catch the Germans in their bunkers. In the meantime, they waited. It was bitterly cold, the coldest winter in twenty years. The rivers had frozen two-feet thick. The mud gleamed with a slippery crust of hoar frost. On the eve of battle, wrote H.R.H. Clyne, the shivering men behind the front lines lit fires to warm themselves: "And I saw on that night Henry V at Agincourt wandering around from little fire to little fire: 'And gentlemen of England now abed will curse the day they were not here.' It was exactly that. That exact scene."

In the tunnel, Henley thought not of Shakespeare but of his objective at La Folie Copse. Overhead, at precisely 5:30 a.m., like the crack of doom, 1,097 guns fired simultaneously. Their shells rustled overhead and the ground above began to shake. Canon F.G. Scott watched from a distance. "The flashes of guns in all directions made lightnings in the dawn. The swish of shells through the air was continuous and far over on the German trenches I saw the burst of flame and smoke in a long continuous line, and above the smoke, the white, red and green lights which were the SOS signals from the terrified enemy." When German batteries returned fire, they were immediately pulverized by McNaughton's heavy artillery. Henley and the Black Watch emerged from the tunnel at the edge of a vast crater. A sliver of dawn was wedged into the horizon but the sky was overcast and it was snowing. Freezing mud stuck to everything. Frost crunched underfoot. "We were right on the heels of our creeping barrage. Some of our men got killed by our own shells. We had seven jumps in the barrage," Henley says. "I don't know any more about the battle except what happened in my field of vision. You can't describe any battle, I don't give a damn who you are. If the guy next to you is hit, even if he's your own brother, you just keep going."

Alexander McClintock described it this way in 1917: "You see two walls of flame in front of you, one where your own barrage is playing, and one where the enemy guns are firing, and you see two more walls of flame behind you, one where the enemy barrage is playing, and one where your own guns are firing. And amid it all you are deafened by titanic explosions which have merged into one roar of thunderous sound, while acrid fumes choke and blind you. To use a fitting, if not original phrase, it's just 'Hell with the lid off.'"

Up this impregnable German fortification the Canadians went and then over the top and down the other side. With them went a plain red silk ensign. It was the flag that had been presented to Lieutenant P.E. Coleman that tawny summer evening of August 11, 1914, when thirty-eight of the first British Columbia volunteers, still in their civilian clothes, paraded at the corner of Nanaimo and Martin streets in Penticton. The flag they carried to the top of Vimy Ridge resides today, almost forgotten, in the city's museum.

"Vimy Ridge was absolutely central to the formation of Canadian nationhood," concurs Reginald Roy, one of Canada's leading historians. "We became a nation there in the eyes of the world.

"From the time the British captured Canada there was a colonial attitude—we were good but everything British was better. Vimy Ridge was where we shed that colonial attitude.

"It was a nation-wide phenomenon. It cut across French and English, rich and poor, urban and rural. Vimy Ridge confirmed that we were as good as, if not better than, any European power."

Henley points to a photograph of Lieutenant-General Currie taking King George V on a tour of the Canadian front. He says that for him the photo has always symbolized the change in attitude. In it, Currie walks the dry duckboard as though he owns it. The British general staff officer beside the king walks in the mud.

This new and powerful self-confidence was forged at Vimy Ridge from an unlikely confederation of 97,000 Canadians from every corner of the dominion. They were prairie farm boys and up-coast loggers, hardrock miners from Trail and sophisticated Montrealers, school teachers from Victoria and nannies from Toronto, Mohawks and Cape Breton coal miners, Inuit hunters and Anglo-Irish horse lords from the arid BC Interior, trappers from the Stikine and drygoods clerks from downtown Vancouver.

In BC, with a population of around 400,000, more than 40,000 enlisted. Little Vancouver Island mustered ten battalions; Greater Vancouver mustered twelve. Across Canada, almost one in six people between the ages of fifteen and fifty-four had volunteered.

Over the next four days at Vimy, these Canadians, led by Canadians, following a meticulous plan devised by Canadian officers, would win a battle that would stun their old colonial masters, refashion the methods of warfare, alter the face of Europe, change the course of history and thrust a new nation onto the world stage. "It is really from that time onward that the Canadians think of themselves as an absolutely elite corps," says Roy, whose father won a Military Medal at Vimy with Calgary's 31st Battalion. "There they stood on Vimy Ridge," said Byng in his post-war tribute to the Canadian Corps. "Men from Quebec stood shoulder to shoulder with men from Ontario, men from the Maritimes with men from British Columbia, and there was forged a nation tempered by the fires of sacrifice and hammered on the anvil of high adventure."

After Vimy, nothing would be the same. Canada would shrug off its notions of British and French superiority. Canadians became the shock troops in the allied offensives that led to victory. After Vimy would come the full flowering of the Group of Seven and Emily Carr, new kinds of literature, the transformation of a sleepy agrarian backwater into a major industrial power. After Vimy would come an independent place for Canada at the Treaty of Versailles, a seat in the League of Nations, full independence from Britain, a sense of national purpose that would prepare it for the next Holocaust nobody believed would come. But it was there, on Vimy Ridge, born of the sufferings and triumphs of its common people, that Canada first imagined itself a complete nation fully capable of asserting itself in world affairs and with every right to do so.

And it is here, in Penticton's modest museum, that you can see the banner that may boast the most remarkable record of any flag flown during the Great War. It accompanied the Canadians through every action in which they took part, with one exception: Passchendaele. It was in hospital with the soldier in charge of it. But it went into No Man's Land on night patrol, right up to the German trenches, wrapped under a soldier's battle dress. It went over the top and through the wire. It went through gas and shrapnel that ate regiments whole. It escaped the disaster at Sanctuary Wood, on June 2, 1916, when the whole company under Lieutenant Colonel Coleman was annihilated. It was carried through the battles at Amiens, Arras, Canal du Nord, Douai, Vimy and Valenciennes and Hill 70. It flew at Ypres and Dickebusch, Ploegsteert and Messines. It was in the battle of the Somme and present when tanks were first used at Cambrai and later at Poziere during the bloody attack on Regina Trench. It was flying at Mons when Armistice was declared. And then it had one last adventure. On November 11, 1918, Lieutenant Colonel Coleman carried the Penticton flag through a cavalry screen still guarding the Belgian frontier and went on into Germany, two days before the authorized crossing of the border. At 1:10 p.m., passing through the village of Maldinger, the tattered rectangle of silk from Penticton became the first Allied flag to fly in victory inside the Kaiser's empire.

Years later, almost as an afterthought, Lieutenant Colonel Coleman returned by mail what had been given to him when he mustered those ill-clad troops at the dusty corner of Nanaimo and Martin in August of 1914, the flag that had flown over Roy Henley in the middle of a snowstorm when a new kind of Canada emerged from all that blood and sorrow.

The Big Cedar

This night comes with a hard frost in it. Dewdrops compress into diamonds. They click and dangle from the tips of the salal. Underfoot, the puddles have evaporated from under shells of cloudy ice. They crunch and crack like rifle shots. All along the skyline, the stars are bright and hard. They glitter like glass beads on a black velvet dress—and they seem to move, as though some supple body were stirring under the silken sky of night. Here, in the harsh latitudes of the province's North Coast, the trees of an austere rain forest root themselves in the peat bogs of Porcher Island and straggle inland towards dry ground. Most of them are scrub, but along the unnamed creeks, they sway and sigh. Their old men's beards of moss whisper against the wind.

How do we value the life of a tree? Board feet, tons of pulp, units to be calculated, stripped and shipped to feed industrial enterprise? Or jobs, since some believe any job is honourable simply because it's a job? Some measure a tree against the works of mankind. They see a century, a millennium, three thousand years of life in trees older than the great cathedrals of Europe, older than the faith they celebrate in Canterbury, Chartres or even Rome.

They see trees reaching toward heaven as those stone saints reach. Some measure a tree only against their own small lives. Memories and dreams: "that tree was there when I was a boy and my sister wore the beauty of youth; it's there still; will it be there for my granddaughter, too?"

Maybe it's just my host's blue-veined hands against the table cloth. Nearing eighty, he turns the pages of photo albums. He is my archaeologist of time, this man who fished the great waters of the north Pacific before the wildness bled out of the world, who watched the schooners go out under sail. Layer by layer, our task illuminated by the sepia glow of a kerosene lamp, we bare the sediments of his heart. Old friends, his long-buried mother, a father who was drowned at sea, schoolmates, the dead, the living, a cherished boat. Until we come to the picture he is searching for.

In it, a beautiful girl and a boy with striking, aquiline features. They lean against one another, shoulder to shoulder, less romantic than loving in the sense of having entered love's landscape before awakening to its limitless dominion—as do we all that first and only time. "My sister."

Like its owner, the tough longliner gone suddenly frail, almost transparent in his years, the picture shows the ravages of time. Silver emulsions drab away into brown under the burden of three-quarters of a century, the edges soft and fraying, a card played one too many times. Only memory remains sharp and hard. "I don't know what happened to that boy. I know she didn't marry him." The gas lamp hisses. The wind roars over the homestead. The forest sighs and, beyond it, the tide rises with the moon and it sighs too, breaking all along that rocky coast. Behind the girl and boy, so young so long ago, a powerful icon of timelessness. A tree so huge it almost fills the picture frame. How many generations of our own muscle and blood, birthing and dying, are spanned by that tree?

The next morning, I go to find the great cedar of the old man's childhood. Upstream to the three forks. Past the big pool. North by northeast toward Salt Lagoon. Two hours into the bush. The tree stood on a sunny slope, immense, perhaps because of better purchase and drier ground. I stood up on the stump where girl and boy leaned. At its base it measured six of my arm spans in circumference—thirty-three feet around. The trunk shattered when it fell, the heart gone punky over centuries. The wood proved useless, I assume, so the loggers left it in the slash to rot. This cut seems hideous, even by the slack standards of clearcut management. Private land, the locals shrug. You can do anything. Craters yawn where the blowdowns have pulled root clusters out of the thin mantle of earth. Whole trees felled into creek cuts and abandoned. Already the debris piles up behind latticed branches. I didn't tell the old man its fate when he asked me if I'd found his talisman. "Yes," I said. "I found that tree."

"If trees have analogies to human families, and I am sure they do, how can we clearcut all their relatives, young and old, not to mention ancestors and descendants, the stock of generations, and expect them to accompany us as useful resources," asks the poet John Hay. Trees are jobs, we answer, secure in the belief that any job is made honourable simply by being a job.

Fire Down Camp

When you live in the oldest house in a community of century-old houses, the yellow flash of the fire chief's van racing past the front porch is always noteworthy. These houses Down Camp have all been altered in some way by the five generations of families that have lived in them since Union was settled in 1888. This one was built when Sir John A. Macdonald was prime minister and Susan bought it for less than the price of a new car when she came here to research her biography on Ginger Goodwin, a labour leader who was shot by the Dominion Police in 1918 and boarded in a house just up at the end of the street. The shingles are painted a rusty barnyard red—clearly somebody got a deal at a paint sale. She stripped the linoleum and found beautiful, honey-coloured fir planks that gave a lovely warm lustre to the 750 square feet that was once thought adequate for families of six or more.

Out back is a kitchen garden full of pot herbs and a small sunny meadow with what are now called heritage apple trees, producing small tart fruit that nobody seems able to identify but which render exquisite pies. Somebody said they are Italian apples, somebody else that they're an ancient variety from Somerset. The fact is that nobody really worries much about their origins, least of all the bears that climb after the ripe fruit each fall. Among my daughter's first words, waving her Pablum-encrusted spoon in excitement, was "Ba." I thought it was a comment on her breakfast but it proved to be directed at the fat little bear she could see through the kitchen window. The apple, pear and cherry trees along with the garden vegetables and the trout, salmon, partridge and deer—they still call a nice four-point buck a "bush salmon" around here—were an integral part of household food supplies in a time when families were large and wages were small and intermittent. And now the fruit has become part of the annual food supply for birds, raccoons and bears, too. One morning after the night before, people walking down the Camp road heard loud snoring coming from the tall grass in the vacant lot next door and stopped to investigate. They found that bear, its round, hairless tummy rising and falling with each breath, its limbs jerking in whatever dreams bears have, sleeping off the effects of a bushel or so of well-fermented windfalls. Then it woke up, saw the semi-circle of curious faces and leaped up. People and bear fled in all directions.

The bears, the apples and the herbs in the old, old garden serve to connect us to a collective communty past that grows hazier with the passing of each generation. Every apple cobbler or piece of salmon dropped off by our neighbour who fishes out of Fair Harbour and baked with herbs picked out back is part of a communion with that past, a new layering of experience into the often unspoken wisdom and unconscious history that textures a community's sense of itself. Replacing the roof on her old house, Susan found that somebody had burned a hole in it years before and, money being short, had just patched it and tacked on a new layer of shingles. She took everything down to the roof rafters and found the remnants of five or six successive roofs up there, a kind of metaphor for the successive lives lived below.

The inhabitants' occupations Down Camp have changed over

the years since coal was king and this community was one of the most prosperous in British Columbia. Muckers, mule drivers and rope riders gave way first to fallers and rigging-slingers, then to supersonic warplane mechanics and computer jockeys from CFB Comox. The evolution from mining camp to bedroom community has been a long and arduous path. Each generation of tenants added personal touches to the eccentric jumble of styles and oddities that distinguishes this end of Cumberland, a village of 2,500 with a mayor affectionately called Bronco by his constituents—everybody here has a nickname—that nestles at the foot of the Beaufort Range beneath Forbidden Plateau and the glaciers and snowfields that feed deep, dark, icy Comox Lake. Over in upscale Comox proper they've been known to refer to Cumberland as "Dodge," as in "time to get outta Dodge." Folklore has it that it can get a little rowdy up here on a Saturday night when the patrons of the Waverly and the King George have had a few more than they should. All I can say is that it seems as quiet as a tomb compared to what goes on every night in downtown Vancouver or Victoria.

Anyhow, there's a legend to keep up. What else should you expect in a mining town where the boys would stop on their way home to Camp to cut the dust at the Bucket of Blood, the Vendome or Shorty's Poolroom? These days Cumberland is reinventing itself from its own past. There's a new cultural and heritage centre adjacent to the impressive little museum, character boutiques and restaurants that cater to a new generation to which all this history seems exotic and mysterious. They troop past the cairns at the long-vanished pitheads, wander through the bramble-filled marshes where Chinatown once stood, make their pilgrimage to the cemetery where a granite boulder engraved with "A Worker's Friend" marks Ginger's grave, and many of the other Victorian tombstones are for those killed in mining disasters.

Camp itself is a living window on that past and how it becomes the living present: a little gingerbread here, a little chainsaw art there, veranda railings to represent each decade's tastes, rock gardens and studio additions, sunrooms and sundecks, dormers and skylights, the odd extra room tacked on here and there. Increasingly, low-maintenance vinyl siding and aluminum windows replace the original stained shingles and wooden frames. Asphalt tiles displace thick cedar shakes. There's even a new house on one of the vacant lots. Down in the cellar of this particular house—the word basement would be a bit too dignified in describing the raw earth covered by two-inch planks that were salvaged from a barge that broke up on one of the Baynes Sound beaches in a winter storm—is a stump that's more than two metres in diameter. It looks pretty much as though the builders, in a hurry to get housing for the new mine, felled the great Douglas fir, milled it on the spot and built the house right over the stump of the tree that provided the wood. At their hearts, all these houses are the same. They are clear, dry, tight-grained timber. Instead of studs and fire retardant insulation, our walls are thick planks as hard as iron and drier after eleven decades than the best kiln-dried lumber. These houses, in short, are built with the kind of timber that a fire may gallop through in the twinkling of an eye. Some Camp folks still remember the uptown fire of 1932. Our neighbourhood old-timers certainly do. Driven by one of those furnace winds that descend mysteriously from the snowfields of Forbidden Plateau, the fire started in the basement of the Campbell Brothers' store and quickly spread to Ruby's Cafe before consuming the Ilo Ilo Theatre, the Masonic Temple and Henderson's Candy Store, all in a matter of minutes. And then there was the Big Fire of 1938 that burned all the way down from Campbell River and threatened to consume all these small communities—Merville, Bevan, Headquarters—many of which survived the fire but vanished when the miners and loggers moved on.

I thought about this when Fire Chief Ken Allen whizzed by and the alarm sounded to call volunteers to the fire station. I'd been working outside to replace the old plank siding above a ground-level wall section we pulled out to get at some foundations which were in serious need of restoration. Roughing in cedar boards, shiplapped in the original style, is harder work than it sounds on a wall with so many add-ons that the planes and angles go every which way. So I was thankful of the diversion. I took a moment to sit on my sawhorse, mop my face and eavesdrop on the domestic confab between Susan, who came out to lean over the veranda rail, and Dot Buchanan, who came out to take her characteristic stance at the roadside, feet splayed, hands firmly on her hips, both of them gossiping about just what was going on down the road past the old apple orchard toward what used to be Chinatown.

About the time the first wisps of smoke began to drift up through the creamy spring blossoms of the long-abandoned cherry orchards at the far end of Camp, the trucks of the volunteer fire department came booming along our narrow road. This made it a serious matter indeed. Susan was delegated by Dot to go get the facts and report back if we had to start limbering up the garden hoses. Off she went, joining delegates from other Camp households down the ragged-pants end of Dunsmuir Avenue. They straggled along to where it peters out, just short of the weathered log cabin that used to belong to Jumbo and which is all that survives from now-vanished Chinatown. Back up the line came the word: "Looks like it's that rented Bevan house set back in the trees."

Once you learn the characteristic roofline—Egyptian pyramid—you recognize Bevan houses all over Vancouver Island. About seventy years ago when Bevan townsite was abandoned, all fifty of the company houses were recycled. Many came up to Cumberland but others seem to have wandered as far afield as Fanny Bay, Bowser, even Duncan and Campbell River. Bevan house or not, this one was at the end of its long life. "She's burning pretty good. The tin roof is holding the fire inside. Nobody's home and nobody's hurt," comes relieved word from our raggle-taggle reconnaissance team, passed up the bush telegraph from porch to porch.

Jo Yacub leans out her window to yell motherly instructions at the fire monitors two doors down: "If there are any kids down there, keep them clear of those firehoses!"

Another woman interrupts a speculation that the fire started with a stove left on—"Gee, hope I didn't leave *my* stove on!"—and hurries home to check. At least the fire trucks are handy.

Soon the watchers trickle away, their enthusiasm dampened by a sharp, sudden rain mixed with hail. The house's tin roof is now peeled open and curling from the heat, reassuring the rest of us that our volunteer firefighters—highly trained and highly regarded for a good century—are in full control. Camp is not at risk, not today. It's time for neighbourly coffee, and the sober bull sessions with Dot and Buck Buchanan and Elsie Coe just up the road. They will take longer to conclude than the old house took to burn.

Fishing on the Blue Line

Storm warnings over the future fate of the salmon fishery now freshen like the westerlies that set the relentless sea pounding this rugged coastline into the sandy beaches that punctuate its rocky headlands. Windjammer skippers called this coast the Graveyard of the Pacific and not without reason. From clipper ships to container vessels with computer-assisted navigation, more than two hundred wrecks are strewn along the reefs, shoals and beaches that make an arc from fog-bound Cape Flattery to wind-scoured Cape Scott. After the SS *Valencia* foundered and 126 drowned off the Klanawa River in 1906, a lifesaving trail was hacked from the bush. It snakes 80 kilometres down Vancouver Island's exposed southwest coast.

If anything symbolizes changing values in British Columbia, perhaps it's the West Coast Trail from Pachena Point to San Juan Harbour. Now an internationally acclaimed tourist attraction, the trail is so popular that the week-long hikes have to be rationed. You're as likely to hear German or Japanese in the Rain Forest Cafe at Port Renfrew as you are a logger's twang. And any laconic chit-chat about skidders and backspars takes place not under girlie calendars from machine parts companies, but beneath the ultra green posters that feature giant trees from the Carmanah Valley and still contentious Clayoquot Sound.

Values are changing at sea, as well. The vast runs of sockeye, spring, coho, chum and pink salmon that once seemed to symbolize the endless fecundity of nature now represent only limits, fragility, human incompetence and the threat of ecological disaster. Despite attempts at intensive management, it's still safe to conclude that the wild bounty of the salmon remains under perilous siege from its old enemies—human greed, stupidity and pig-headed intransigence. The west coast of Vancouver Island is where some of the bloodiest skirmishes in the five-year fish-war between Canada and the United States were fought. Here, on what's known as the Blue Line, Canadian boats would intercept both sockeye bound for the Fraser River and coho and chinook bound for rivers in Washington and Oregon.

In 1993, I drove west from the smug suburbs of the provincial capital at Victoria for a visit to the front lines. My route took me 70 kilometres, out past the fishing port of Sooke, colonized as a bedroom community, past the dryland sorts of Jordan River logging camps, past China Beach and the Sombrio River. The gentle, park-like setting of south Vancouver Island gives way to rugged highlands lashed by heavy weather from the open ocean and the Japan Current. The narrow road dwindles to a single lane for Bailey bridges shared with loaded logging trucks. It winds through re-forested clearcuts and active logging shows before twisting down to Port Renfrew. At the end of the road is a single dock. It is suspended on pilings that stand high above tidewater to accommodate the big tides and winter surges that sweep in from the Swiftsure Bank. Tied up at the dock, invisible except for a glimpse of her masts and rigging, I found John Chorney's 40-foot steel-hulled gillnetter, the *Gala Babe II.* John hails from Delta and the mouth of the Fraser. He was busy wiping acidic road dust off Oline

Luinenburg's survival suit before stowing it in the cabin.

Oline is the official observer from the Pacific Salmon Commission. She drove the axle-busting back roads from her home at Cowichan Lake just to avoid the commuter traffic around Victoria. "I didn't know there were so many holes in my car," she tells me over coffee. "Car's full of dust. Lost a muffler, I think. I bottomed out a lot on the way through." We're killing time while John and his son John Jr. check gear in preparation for sailing. His boat is contracted by the commission for a series of test sets out in the windswept entrance to Juan de Fuca Strait and I'm going with them.

We'd drunk the coffee pot dry at the hotel talking about Oline's master's thesis at Simon Fraser University. It's two weeks from completion but her hopes of squeezing enough time out of the job to wrap it up seemed dim. From tonight Oline will spend the next six weeks working fourteen hours a day, seven days a week, on the rolling, heaving deck of a fishboat in Area 20, as the entrance to the Strait of Juan de Fuca is designated in federal fisheries regulations. Twelve nautical miles away is the American shore. The boundary runs right down the middle of the strait. But 17 million salmon don't care about borders. They go where nature tells them. Thus the feverish intensity of finding out where that will be. Canadians have a huge investment in these fish. The taxpayer has spent billions to protect, expand and manage the resource. High-value BC sockeye stocks are an environmental success story. Because we've foregone the quick payoffs of hydro-electric development on the Fraser, invested in habitat protection and volunteered millions of private citizens' hours for enhancement projects, this sustainable resource is actually expanding. The Americans, on the other hand, have done a terrible job. They ravaged the salmon resource to the extent that major closures are in effect from Washington to California. The Columbia, once a salmon factory to match the Fraser, has been ruined with dams and diversions. Some

American stocks have been declared endangered species. It's not surprising that the idea of American boats plundering fish that Canadians have nurtured and paid for offends almost everyone. Each Fraser River sockeye in an American hold denies the Canadian taxpayer a return on the investment. Mind you, the idea of Canadians fishing hard on American stocks that teeter at the brink of extinction is no less offensive.

Oline's job does not involve ideology. It involves counting the species caught, recording water temperatures and taking scale samples for analysis by the Pacific Salmon Commission laboratory, which lurks above an upscale law office among the trendy boutiques of Robson Street in downtown Vancouver. As with tree rings, each scale writes the life story of a fish for the scientists and technicians charged with determining the real numbers and abundance of the Fraser-bound sockeye. And like human fingerprints, scales can trace each fish back to the gravel bed where it hatched. Adams River sockeye form a race as distinct from Early Stuarts as Caucasians are from Asians. There are twenty of these categories—Horsefly, Chilko, Stellako, Cultus, and so on—to be identified and conditions of their return journey to the spawning grounds recorded. On the basis of what the Pacific Salmon Commission determines, catch allotments are doled out: so many fish for the spawning escapement; so many for the outside trollers; so many for small boats in the gentler waters behind Vancouver Island; so many for Juan de Fuca purse seiners; so many for gillnetters like John Chorney; so many for the aboriginal fishery; so many for the Americans.

All this seems a far cry from academic scholarship. What's a writer of fiction, poetry and feminist essays—she says her thesis is a deconstruction of the language used in attempts to educate women about safe sex in the age of AIDS—doing up to her knees in slime and dogfish? Oline says she's out here counting salmon on a six-month contract because she loves the sea, can't stay away from it. She's already sailed a yacht to New Zealand, crewing for a friend, so she's well aware of the dangers from big Mother Ocean. Last summer on a test set out of Winter Harbour on the north end of Vancouver Island there were moments when she wondered about her wisdom. "We got caught in a winter storm that happened in the middle of summer," she says with a wince. "We were the only boat out there. It was terrifying. Nobody thought we were getting back in. We spent a lot of time with our feet in the immersion suits."

I felt a little bit of the same as I contemplated the swinging chain ladder we'd have to negotiate to reach the moving deck of the *Gala Babe II* a couple of storeys below. I have no fear of heights, but I'm long past the agility of my youth. A ladder dangling in the wind above a long drop into the chuck is not a welcome sight. My guide swarmed down with the youthful skill of a gymnast and there was nothing for it but to follow. John Jr. took one look and got the boathook out just in case, an act of generosity which did little to instill confidence. But I made it aboard, boathook not necessary.

It isn't long before the vast power of the Pacific announces its presence. The boat's frame shudders each time we nose into a swell. Windshield wipers tick rhythmically, sweeping away sheets of spray that lash the foredeck. The skipper reassures me. He has 700 horsepower at his disposal in the twin diesels below decks. She's just idling, he grins. If we want to go—or have to—she's a race-horse. Then he tells me I've chosen a beautiful night for my first look at high seas gillnetting. John should know. He's been fishing off the west coast for forty-three years. He began as a fourteen-year-old setting nets from a rowboat in the mouth of the Fraser. By now he's fished everything—halibut, herring, salmon—from Alaska to Washington State. He's good, too, consistently good, which is why he's under contract to run tests for the Pacific Salmon Commission as estimates of returning salmon are calculated.

What does he think for the fishery after nearly half a century of watching it? Too many fingers in the same pie. Everybody thinks their slice should be bigger than the next guy's. But there's only one pie. You can't bake another one. What's the solution? Try to remove self-interest from any negotiations over who gets to catch what, he says. Self-interest diverts discussion from the key issue—the only issue—which is the long-term well-being of the fish.

No fish, then nothing for any of us to discuss, right? He thinks hard times lie ahead for the fishery. "I'll be okay. I'll survive, but I feel for young guys like him." He nods at John Jr., shrugging into foul weather gear in preparation for the first set.

Tonight we're bound to fish a precise location in 80 fathoms about 2.5 nautical miles off San Juan Point, well inside the American boundary and right on the migration route for the prized Early Stuart sockeye bound for the Fraser. It was right around this day in 1789 that Jose Narvaez charted this stretch of coastline for Spain using sextant, chronometer and soundings from a plumb line. Guns from an unknown Spanish wreck have been recovered from Koitlah Point on the American side. Today, computerized navigation equipment finds the precise map co-ordinates, allowing for the drift of wind, tide and current. In fact, the eerie glow from high-tech instruments—peach, green, blue from the echo sounder, red from the radiophone, pulsing amber from radars, gold from the Loran—leaves me feeling like I'm in a jet fighter rather than a utilitarian steel fishboat.

Near midnight a pale, oyster-coloured wedge lingers above the black bulk of Vancouver Island's mountains. The heaving sea is the colour of ink that flashes with phosphorescence. The Milky Way is a luminous river above. Under the glare of halogen floods, the two men feed 300 fathoms of net over the cramped stern. It's awkward, bone-chilling work. The summer air may be warm on shore but out here it's as cold as the bottom of a well in February. Even four decades of sea legs stagger and scramble for purchase on the slick, pitching deck. If night makes the mesh invisible to fish, here in the sea lanes we want to know where it is. Bright orange floats and radar reflectors on long poles are clipped to the net as it goes out. Coal-oil storm lanterns on floating platforms mark the ends. Then the net drifts, a glimmering necklace of white floats strung between points of light on the dark sea. Back in the cabin, we drink hot, sweet tea and munch Girl Guide Cookies—John's a big supporter, so there's a whole case aboard—then dim the lights and doze fitfully while John Jr. waches the radar, tracking four huge freighters.

In the wee hours, we're back on deck to retrieve. The net is laden with mackerel brought north by the warm current called El Niño. There are many hake, some skipjack tuna and dogfish with razor-blade mouths and poisonous barbs. A power winch pulls in the net, but the by-catch must still be removed piece by wet, slippery piece. Oline crouches, counting each fish in the net and recording it in her log. John estimates 150 sockeye would pay costs—"I've had 1,400 sockeye on a single set!"—but the net is two-thirds in before we find the first salmon among the other species, a big, silver torpedo with its characteristic metallic green head. It goes into crushed ice in the hold. A few others follow, but not many. He pays the nets out again for a second set then retires to the cabin for more tea. I drift off, asleep in the arms of the slow swinging sea.

At first light we pull in nets again. By 8:30 a.m. we're back in Port Renfrew. A pair of crab fishers gratefully load more than six hundred pieces of the by-catch into their skiff for trap bait. On the dock, Oline makes her final tally: a few small coho, a couple of steelhead, some large springs, fewer than twenty sockeye. In all, about $100 worth of salmon. John shakes his head. The night's fishing used about $200 worth of fuel, not to mention the labour. It's a reminder that for all its images of robust self-reliance, a fragile way of life hangs in the balance as we struggle to manage this miracle, this incredible gift from nature: the salmon.

HUCKLEBERRIES

In the season of the huckleberry, the bush being laden with the plumpest yet seen, the rotten stump being low, what else is there to do but stop in the rain and genuflect before the high priest of wild fruit here on the North Island? On all sides, the massive trunks rise like pillars in an Egyptian temple. The jade-green forest steps away into a sombre labyrinth that sweeps all the way from San Josef Bay to the mouth of the Sushartie, connected only by the rotting corduroy trails laid down by the inhabitants of long-vanished settlements. Behind every tree-framed chamber in the gloom, another chamber is glimpsed, and then another, then another, each opening different, each the same, until the seen and the unseen are confused in a paralyzing complexity of light and shadow.

I am on the west coast of Vancouver Island, at least 80 kilometres beyond the Island Highway that reaches back from Port Hardy to Victoria. To get here, I followed the rough gravel, took my Jeep down an ancient dirt track that petered out in a borrow pit, then hiked another half day. Now the only sound in the stillness is the patter of a fine rain. It is punctuated by the resonant plop of larger droplets. They gather along branches in the treetops before plunging into the understorey to make the salal dance. In a hollow, the luminous glimmer of skunk cabbage. The pungent censers steam below streamers of moss. And then, in an opening where light penetrates the high canopy, the decaying stump and its acid-green corona of *vaccinium parvifolium Smith*.

For me, at this time of year, all of British Columbia's rich diversity can be distilled into one sensation—the sweetly sharp bite of a huckleberry crunched between the teeth. It comes with the inescapable sense that something slightly wicked attends that hint of sweetness laying ambush in the tart juice. Perhaps it is just the fact that the red huckleberry has so far escaped domestication—at least, I'm not aware of any commercial huckleberry farmers diligently striving to breed out the wild eccentricities in favour of uniform size and sweetness sufficient for the saturated urban palate. So the huckleberry remains a leader in the wild resistance. It takes root where it can and every encounter is a serendipity. You have to go into nature to get your huckleberry. In this alone, it reminds us of ancient origins in which our own survival thrived on nature's untended bounty.

Sometimes it's called the red bilberry. Whatever nomenclature you prefer, it remains one of the most common of our wild berries here in what poet Charles Lillard liked to call the Sitka biome, this place on the wet side of the mountains where rain and rot and fecundity seem to seep into the very culture itself. Abundant from Vancouver Island to Alaska and in parts of the southern interior, this deciduous shrub prefers the deep shade of the understorey. It's often found growing in nurse logs—the rotting remnants of huge blowdowns—and the decaying stumps left by loggers or mother nature herself. These berries were a staple of the coastal peoples before contact and they remain an important component in what's sometimes called "country food." In any event, large quantities are still eaten today.

When fully ripe, they can be harvested by shaking the berry-laden branches over a blanket. Sometimes special combs are still used to rake the berries off the bush and into a bark box or basket that hangs under it. Kwakwaka'wakw women clean the berries of their twigs and leaves by rolling them down a wet board. While the clean berries roll to the bottom, the twigs and leaves stick. The berries would be boiled in high cedar boxes, the water heated with stones from a firepit, then mixed in red salmon roe, covered with hot skunk cabbage leaves and sealed in the box with the grease rendered from eulachons. They'd be eaten as a feast delicacy during winter ceremonials. The Nuu-chah-nulth, on the other hand, eat huckleberries fresh or not at all. Only thimbleberries or the dark blue berries of the salal would be preserved, usually in dried cakes. But among the Sto:lo and Nuxalk, they were preferred mashed and preserved as dried cakes.

If there are regional differences regarding the best way to consume the huckleberry, identifying the exact berry to which the term refers is one of the great divides that distinguishes coast people from those who inhabit the rest of the continent. There are about a dozen related shrubs in the genus *Vaccinium*. Salal berries, kinnikinnick, blueberries, bilberries, whortleberries, cranberries are all called huckleberries in some parts of North America. Praise the red huckleberry on the other side of the Coast Range and you're likely to earn contempt from the dry-landers who think it insipid by comparison to the meaty dark blue berries that carry the name on their side of the mountains. East of the great divide, huckleberry is generally taken to mean the fruit of genus *Gaylussacia*. These berries come with ten hard seed-like nutlets inside the pulp. They do make wonderful jam. And this is the fruit that kept Huckleberry Finn while he was on the lam.

It turns out there's no right or wrong in this debate over nomenclature. The etymology of the huckleberry is unclear. The root word, *huckle*, is an old word of obscure origins which means to barter or bargain. Perhaps huckleberry derives from the early

Elizabethan settlers' encounter with dried berry cakes as a trade item during their first hard winters on the eastern seaboard. On the other hand, there's a suggestion that huckleberry might simply be an illiterate pioneer's corruption of the word whortleberry, which is also part of the same family.

But here on the Rain Coast, the huckleberry is not blue and doesn't grow on some low-lying shrub. For us the huckleberry is the red one that looks similar to a blueberry and grows proudly on bushes that sometimes hang nicely over your head. I'm told that huckleberries make a tolerably good wine but that you have to work a little harder to start the fermentation process. That, I've never tried. And I'm told that huckleberries are easily preserved by freezing—one way is to scatter them one layer deep on sheets of aluminum foil and lay them flat. But don't add sugar first, it toughens the fruit.

There are other indigenous berries of note, of course. You can't beat a bowl of ripe salmonberries with thick cream, and I don't care what the nutrition police have to say about the hazards of milk fat. Even without the cream, these bright yellow and orange and sometimes red berries have saved the day for more than one angler who dumped the packet of sandwiches in the river while opening the creel to put in a trout. In fact, the fruit of *Rubus spectabilis Pursh* is called salmonberry in acknowledgement of its resemblance to the roe clusters that appear from high summer to late fall in the rivers beside which the berries grow in dense thickets of canes and rustling leaves. The tender shoots of salmonberry were of great value to aboriginal people. Emerging early in the hungry time between the end of winter preserves and the onset of summer fruit and the first salmon runs, they are "muckamuck," eaten fresh or roasted. Close to the salmonberry is *Rubus parviflorus Nutt*, the thimbleberry, the fleshy, dome-shaped fruit that goes squishy when ripe and is common along roadsides and riverbanks wherever light penetrates the margins of the coastal rain forest. The big, rustling leaves resemble those of the coast's ubiquitous broadleaf maple.

Thimbleberry, too, was important to aboriginal peoples of the coast. Gilbert Malcolm Sproat, who travelled down the Alberni Canal to visit the Nuu-chah-nulth in 1868 was amazed to see large canoes laden to the gunwales with thimbleberry shoots in the early spring. And there's *rubus ursinus*, the sweet little Pacific blackberry, which leaves the larger Himalayan variety tasting vaguely synthetic. Brought to the coast by settlers, the Asian blackberries soon escaped and vast tangles of brambles now boil from the ground wherever there's an opportunity. There's no doubt for those who've gone to the trouble to pick enough of the little native black caps, however, that they are matchless in the pastry department—except for hot huckleberry pie, fresh from the oven, with a scoop of vanilla ice cream on the side.

To bake your huckleberry pie—and you won't get one any other way—you start with your standard pastry recipe. Otherwise, in the spartan conditions of camp, try this pastry substitute passed along by a north woods trapper. Combine about 2/3 of a cup of cooking oil with three tablespoons of milk, pour it over two cups of flour and a teaspoon of salt. Mix it up thoroughly and roll or knead it out. Press the oil pastry into a well-greased pan but keep the pastry a little thicker than you normally would. In a saucepan, melt a quarter cup of butter, add one cup of sugar and stir in four cups of fresh huckleberries. Mix them well but immediately pour the mixture into your pie shell. Cover it with an upper crust if you have an oven. Make sure you cut generous vents in the upper crust. Bake in a preheated oven, moderately hot, for about 50 minutes, or until the crust is golden brown. If you don't have an oven, leave off the top shell. Pay attention! You don't want to burn your shell and it's easy to do over a campfire. You'll be eating this with a spoon. Accompany it with heavy cream, whipped cream or vanilla ice cream.

If pie is beyond you, here's a camp treat you can save for breakfast. To make huckleberry bannock you'll need two cups of flour, three teaspoons of baking soda, half a teaspoon of salt, six tablespoons of margarine or butter and a quarter cup of powdered milk. Mix together the dry ingredients, cut in the butter. Add a cup of washed huckleberries, still damp. Mix in just enough cold water to make a dough, about a third of a cup. Don't knead it—just stir it to an even consistency and then shape into a cake about an inch thick. Lay your cake into a well-greased, already warm skillet. Hold it over the fire until the bottom just forms a crust. You'll need to jiggle the skillet a little to keep it from sticking. When the bannock is cooked enough on the bottom to lift, turn it by putting a plate over the skillet and inverting it. Cook until the second crust forms. Stick a fish-filleting knife into the dough to see if it's cooked. If the blade comes out with dough on it, the bannock's not done. Alternately, if this seems too much work, wrap your bannock around sticks and prop it near the fire to cook. Either way, eat it with plenty of butter.

Huckleberries can add some zing to your main course, too. Here's a camp recipe for salmon although it works with just about any fish. Fry a couple of big onions in butter or olive oil in a skillet with a lid. If you don't have a lid, a piece of cedar board will do. When the onions are transparent, throw in a generous handful or two of whatever huckleberries or whatever other berries you've been able to scrounge. Dried cranberries make a good substitute for town cooks.

Sauté the onions and berries for a few minutes, not long. Take out the onions and berries and set them aside. Lay a couple of slabs of salmon or other fish in the skillet, skin side down. Put carrot fingers, zucchini or whatever else is handy and seems like a good idea beside the fish. Cover the whole works with thinly sliced potato—sweet potato or yam is even better—then bury everything with the onion-berry mixture. Put it back on the heat, put on the lid. Let it cook. You won't need to time this. Your sense of smell will tell you when it's done.

Kismet on the Coast

History as we learn it in the classroom often seems distant and dead, a compilation of events involving great men, occasionally women, almost never the kind of people we know. So here's a story that reminds us of how history lives in all of us, how it intrudes into the present to entangle lives and connect them to the past, how love and human passion pulse through all those dry dates and dusty documents. It's a complicated story that jumps backward and forward in time, that involves the wisdom of dead men and the respect of living sons, that reaches from the desk of long-dead United States President Teddy Roosevelt into the office of British Columbia's Premier Glen Clark. It's populated by communists and capitalists and the sweep of past wars between great powers and the present disputes over salmon that embroil Canada and the United States, BC and Alaska.

Let's start the story in Vancouver with the flash of a photographer's camera on Granville Street just outside the Orpheum. It's an evening in the early 1940s and it's a Foncie photo—one of those grainy, black-and-white pictures snapped on the sidewalk to be reclaimed later by anyone who cared to pay and could produce the card handed out by the photographer. The subjects of the photo are a recently-commissioned naval officer and a beautiful young woman with her arm looped through his. He's on leave and they're out for an evening on the town. They stride confidently toward the camera. Above them, the theatre marquees are bright, so maybe it's 1940 or 1941, before the fearful blackouts that followed Pearl Harbor in 1942.

Now we have to jump ahead forty-five years or so to 1985 and a rangy graduate of the Vancouver School of Art named Frank Lewis. Frank is already famous for his murals, the great urban wall paintings that put Chemainus on the map and a 75-square-metre painting for the office of the Prime Minister of Malaysia. This time he's poking around in the archives doing research for a major piece commissioned by the Vancouver Maritime Museum to commemorate the seventy-fifth anniversary of the Royal Canadian Navy. The painting, it's been decided, will honour the largely forgotten Gumboot Navy, the lightly-armed seiners, offshore trollers and pleasure yachts that were crewed by commercial fishermen and patrolled Canada's west coast during World War II. The Fishermen's Reserve had been formed in 1938 and when war with Japan broke out, with the regular navy fighting for its life in the battle of the Atlantic, its seventeen little boats comprised the only naval force available to defend Canada's west coast. But it had its own advantages. Its crews and their skippers knew the coast like the backs of their hands. They knew every reef and shoal, every tiderip and overfall. They had eyes and ears in every Native outport and A-frame camp. They could patrol in fog and heavy weather, hide out in hidden coves and ride the treacherous tides through the islands and inlets.

Looking into the stories of citizen warships with names like *Nenamook* and *Merry Chase, Western Maid* and *BC Lady, Canfisco* and *Springtime V*, Frank comes across the Foncie photo. He puts it aside. It's just one like tens of thousands carried into battle by

Canadian sailors and soldiers and airmen to remember wives and girlfriends and the brief encounters they hope might turn into something better if they ever make it home. But there's something about this photo. Maybe it's the sailor's jaunty air. Maybe it's the smile on the beautiful young woman's face. It's just—something. So Frank works the photo into his painting to represent all those men and women pulled apart and thrown unexpectedly together on the tides of war. His eight-foot collage of images evokes the salty ordinariness of the Gumboot Navy, how it struck to the heart of what it is to be Canadian and from the west coast, "just regular guys going about their jobs as they would have done on the fishboats, except that now they were in uniform." And that's part of the painting, too. The weird bits and pieces of gear scratched together from cadet halls and reserve unit broom closets and boxes bound for rummage sales. The hats with names on the front that bear no relation to any ship. The brash, cocky young men in their hand-me-down bell-bottoms. Frank's painting went on display during Expo '86 as part of the Royal Canadian Academy of Maritime Artists Exhibition and then it went back into the artist's collection in Victoria.

This is where retired naval veteran Arthur Knight enters the story. His father was one of those patriotic British remittance men who had come to the west coast at the turn of the century, went back for the Great War with Princess Patricia's Light Infantry, got blown up and gassed at the first battle of Ypres and came home blinded and suffering from wounds that would plague him for the rest of his life. Arthur was born in Britain while his father was there learning braille but he grew up in Vancouver and enlisted at HMCS *Discovery* himself at the age of seventeen, serving on HMCS *Antigonish, Sorel* and *Crusader* during World War II. He lives in Victoria now, but civilian life took him all over the BC coast and when he saw Frank's painting and what it evoked, he loved it so much that he persuaded the Royal Canadian Naval Veteran's Association to raise $5,000 and buy it. The RCNVA, in turn, loaned the painting out. It went to hang in the veteran's wing of Tillicum Lodge, a seniors' care facility.

Which brings us back to the Foncie photo. "One day at the lodge this old fellow in a wheelchair sees the painting and sings out—'My God, that's my *wife*!'" Knight remembers. The man in the wheelchair, the same jaunty young sailor in the photograph taken during his salad days, is Elgin Neish, better known as Scotty to friends from Coal Harbour to Cape Muzon. The beautiful young woman is his late wife Gladys. Scotty Neish is one of the enduring legends of coast history, a tough, far-sighted man who never compromised his principles. He was fighting against destructive forest practices to protect watersheds and salmon habitat as early as 1939. And he was a life-long labour organizer who was one of the founders of the United Fishermen and Allied Workers' Union. At twenty-three, already skipper of the seiner *Chief Takush*, he led a sixty-boat flotilla into Vancouver to tie up in 1938. That job action won BC fishermen their first collective agreement. Scotty joined the Fishermen's Naval Reserve on September 6, 1939, four days before war was declared and promptly went to sea with the Gumboot Navy as a lieutenant. "We'd get a call that there was an unidentified vessel 200 miles off shore," he later recalled. "We were ordered to 'Go investigate' and off we went, even though the biggest gun we had on board was a .45 calibre revolver." Scotty and Arthur with their different naval backgrounds hit it off from the start and when Scotty's son Kevin asked for a photo from the Frank Lewis painting, Arthur got him one.

Which is how Arthur, all spit and polish in his navy blazer, wound up feeling welcome but a bit out of place at Scotty's rough-and-ready working class wake at the clubhouse of his son Steve's rugby team when the old warrior died in 1995. "There must have been 300 people there," Arthur remembers. "The place was just jammed. There were people from Namu and Prince Rupert and Bella Bella—you name a place on this coast and there was somebody from there." Scotty's wake is what brings us to Teddy Roosevelt and Premier Clark and the living link between diplomatic incidents in the last century and diplomatic incidents today. Laid out on a long table was a display of the memorabilia Scotty had collected during his days as a fisherman, labour organizer, ardent leftist and loyal officer of the Gumboot Navy.

Arthur surveyed the books and photographs and documents. What caught his attention was a series of large volumes. "I opened them. There were two volumes of naval charts. Some of them were Russian charts that went back to 1824. There were seven volumes of transcriptions from the actual meetings that determined the A-B Line during the Alaska boundary dispute." The matter of where the boundary between BC and Alaska should lie had smouldered since the United States purchased the territory from Russia in 1867. But it flared up when gold was discovered in the Klondike in 1896 because the Americans wanted freight and passengers bound for the Yukon to pass through their territory and pay their tariffs. A belligerent Teddy Roosevelt, made confident by his victory in the Spanish-American War, pressed the British, whose confidence had been shaken by setbacks of the Boer War, for advantageous terms. When the British mostly acceded to American claims in 1905, Canadian judges refused to sign the accord and the country erupted in an anti-British fury. Some historians suggest that Prime Minister Sir Wilfrid Laurier's inaction resulted in the defeat of his government and its free trade agenda in the Reciprocity Election of 1911. That dispute simmers on today with fishermen on both sides getting arrested and disagreements over precisely where the line coming out of Portland Canal and heading westward through the rich fishing grounds of Dixon Entrance should lie. American and Canadian maps tend to show different boundaries, with the line farther north on Canadian maps and angling farther south on American maps. Which is why Arthur's eyes bugged out when he opened Scotty's transcripts to the pages which showed the A-B Line.

"They were signed by [American senator] Henry Cabot Lodge," he says. "These are the verbatim records of the meetings that took place between the Americans and the British in the last century. And they clearly show the A-B Line running straight out and cutting through the end of Cape Muzon [as Canadians have always claimed].

"I thought, by golly, I pick up the paper every day and there's the premier in some dispute or other with the Alaskans. These papers look important. This is physical proof that the A-B Line runs where we say it does. It's like finding a copy of the Magna Carta in your grandfather's trunk."

So, with permission from Steve and Kevin Neish, he called the premier's office. The premier's office was intrigued, and then just plain excited. "We went to the archives. There is nothing like this in its original form," confirmed Clark's ministerial assistant, Rey Umlas, who examined the transcripts and charts. "The significance of this document is that it establishes that our claim is right. They can't deny it. We have to use this. Mr. Knight, he's a true Canadian and British Columbian to bring this forward. And the Neish family, we owe it all to them."

Arthur thinks the credit belongs only to the foresight of Scotty, who knew the importance of history in shaping the present. "So many people who do real things, their names slip away into oblivion. A hell of a lot of guys did a hell of a lot of things on this coast and we owe it to them to remember that. Personally, I think there should be a Scotty Neish Foundation. There are a hell of a lot of fishermen out there who owe him, who owe him big time."

In the meantime, the premier's office studies the remarkable boundary dispute documents and ponders how to make use of them in the turbulent political debates of the present. And all because an unknown photographer snapped a picture of a passing couple on Granville Street more than half a century ago, and an artist saw something compelling in their faces.

Ocean Plunder

The western edge of Canada meets the boundless Pacific in a foam-laced fracture zone of marine canyons, submerged uplands, towering seamounts and ocean currents that sweep like huge hidden rivers past 29,500 kilometres of coastline and a 6,500-island archipelago. Deep, dark fiords 150 kilometres long bring the pulse of great Mother Ocean right into the heart of British Columbia and the land embraces her with the teeming alluvial fans of more than one hundred great river estuaries and countless streams and brooks that don't even appear on the maps. And yet the greatest richness and diversity of life is not on land, but out there, between and beyond the islands that shield the inner coast from winter storms of hurricane force, where the continental shelf skirts a 2,500-metre-deep abyss known as the Cascadia Basin.

Beneath the spindrift and sea shine, invisible to all but the lucky few of us, sprawls a fantastic underwater biosphere of forests far richer and more diverse than those that blanket the land. Magical, multi-hued groves of tree-like corals, intricate as the embroidered filigree on Elizabethan gowns, ascend in ranks from the depths. Among them, the bottom is vivid with anemones and urchins. Gigantic halibut and supple, dark-skinned sablefish scud like phantoms across lightless bottoms. Far above them, jellyfish pulse through the shafts of light and shadow like the ghosts of countless moons. Schools of needlefish and herring glitter like sudden showers of silver beneath the shimmering, undulating surface film. Immense kelp forests, their long foliage streaming like mermaid's hair, sway in the tidal jets that carry nutrients which sustain a vast diversity of life. Eelgrass undulates across shallower sandy bottoms where crabs and sand dabs and starry flounder lie concealed among shellfish colonies whose oldest citizens were already old when Captain George Vancouver charted these shores.

The BC coast is home to more than 6,500 marine species, including 400 fish, 161 marine birds, twenty-nine mammals, one of the world's largest killer whale populations, nesting grounds for 80 percent of the world's Cassin's auklets and wintering grounds for up to 90 percent of the world's Barrow's goldeneye. And this is just a portion of Canada's coastal engagement with the world's oceans. The Pacific, Arctic and Atlantic rims amount to 243,000 kilometres of coastline. Another 9,500 kilometres surround the Great Lakes. Together they represent the longest national coastline in the world and make Canadians collectively responsible for a vast global inventory of marine life.

Yet throughout this seascape, Canada's delicate underwater ecosystems are routinely subjected to what scientists describe as the submarine equivalent of clearcut logging in sensitive, fragile habitats that might never recover from their destruction. It's done in the name of jobs, in the name of sustaining coastal communities, in the name of feeding a hungry world, in the name of almost every emotive claim that can divert attention from the crude, blundering rape of underwater ecosystems that it represents. Bottom trawling is widely condemned among marine biologists as one of the most destructive forms of fishing. And with fish stocks declining worldwide, scientists have begun calling for a rapid extension of the philosophy that created land-based parks to undersea regions. The question is if it's too late in many places.

Scientists like Dr. Daniel Pauly of the University of BC's Fisheries Centre, writing in the journal *Fish Biology and Fisheries*, say we must begin creating underwater refuges for hard-pressed fish species where no commercial or recreational exploitation of any species should take place. Last year, Dr. Jane Lubchenco, the president of the American Association for the Advancement of Science, led an appeal by 1,600 scientists who want to see at least 20 percent of the world's oceans turned into natural marine reserves by the year 2020. Some environmentalists reject that number as arbitrary and far too conservative. As with everything, there's debate about this. Some argue that setting aside marine protected areas where ecologies can be protected gives a kind of permission for the continuation of destructive practices outside those boundaries, institutionalizing the fragmentation of ecosystems that pay no attention to human maps. Canada, on the other hand, has already been getting poor marks on the World Wildlife Fund's report card, which grades how nations around the world are faring at protecting endangered species, including those beneath the sea. One of the WWF's main complaints was a lack of action by Ottawa on the creation of marine protected areas.

The International Year of the Ocean has come and gone and a new Marine Conservation Areas Act that's been more than a decade in the making was to bring Canada in line with growing international attempts to protect ocean ecosystems. Conceptually, these national marine conservation areas are to be underwater areas managed for sustainable use with smaller zones of higher protection encompassed within them. The areas would include the seabed, its subsoil and the water column above and can be extended to include the transition zone between land and sea represented by wetlands, river estuaries, islands and significant coastal foreshore. These areas are to be managed, says Ottawa—using the new buzzword of the corporate lobby—on a "partnership" basis, the intention being to encourage Canadians to become active stewards of their marine heritage and "to work together towards the common goal of maintaining the area's ecological integrity and ensuring its long-term sustainability."

This country's first national marine conservation area was actually established in Georgian Bay at the north end of Lake Huron in 1987. But in the summer of 1988, still wincing from criticism of its efforts, Ottawa announced that a 3,000-square-kilometre underwater canyon off Nova Scotia, a seamount and thermal vents off the West Coast and 3,050 square kilometres around the Gwaii Haanas National Park Reserve in the Queen Charlotte Islands would also be protected. Work has begun on similar plans for the southern Gulf Islands that lie between Victoria and the Lower Mainland and Johnstone Strait, where Pacific tides funnel into the narrows between the north end of Vancouver Island and the mainland.

What alarms scientists most is the possibility that bottom trawling could be authorized in protected areas considered important enough to be designated for protection. The groundfish and shrimp fishery off the BC coast employs 259 bottom trawl vessels and coupled with shellfish harvesting they generate a landed value of close to $300 million a year, a significant share of the $4 billion a year the marine environment is estimated to contribute to the coast's economy. In one month alone, for example, the commercial fleet landed 3,747 tonnes of groundfish including such diverse species as Yellowtail rockfish, Pacific ocean perch, Dover sole, Rock sole, Rougheye rockfish, Pacific cod, Yellowmouth, Lemon sole, skate, ling cod, pollock and sablefish. The shrimp fishery, which began trawling in the 1960s, landed a record 7,386 tonnes in 1996, largely, the federal government acknowledges, because vessel operators directed increased efforts to grounds with little previous fishing history. In other words, catches are increasing not because the stocks are necessarily healthy, but because technological improvements permit trawl fisheries to operate in previously unexploited submarine terrain. This effort to maintain catch levels that would otherwise decline by expanding the target areas is precisely what disturbs many scientists alarmed by the declining diversity of life in the sea. Some environmentalists have likened the destructive impact of bottom trawling on marine habitats to trying to round up stray dogs by stringing a logging cable and a net between two bulldozers and then driving them back and forth through suburban residential neighbourhoods. "Bottom trawling is simply not compatible with preserving biodiversity in marine environments says Dr. Elliott Norse, president of the Marine Conservation Biology Insititute which is based in Redmond, Washington. "If we want to conserve our fisheries and biological diversity in the ocean then we had better think hard about this."

Sabine Jessen, Vancouver-based executive director of the BC arm of the Canadian Parks and Wilderness Society, says there is growing international consensus among marine biologists that the fishing method not only decimates bottom habitat in the short term, but also destroys its capacity for regeneration in the long term. The society generally supports Bill C-48 but remains deeply concerned about the omission of a statutory prohibition on bottom trawling in waters set aside as marine conservation areas. "You really have to wonder what they are thinking about," Jessen says. "International scientists are calling for a ban on bottom trawl fisheries. There are suggestions that the reason the northern cod are not coming back is because bottom trawling has left no habitat for juvenile cod. We're quite disappointed that the federal government is not prepared to prohibit bottom trawl fisheries from these new marine conservation areas. It's not at all compatible with their ecological aims." The World Wildlife Fund agrees. "An activity with impacts analogous to forest clearcutting does not belong in areas set aside to protect and preserve natural ecosystems," it said in a recent briefing paper.

Bottom trawling is a fishing method in which large, powerful vessels drag huge pouch-like nets across the bottom to catch deep swimming fish and shellfish. In trawling, a heavily weighted cable called a footrope holds the net to the bottom. The footrope can be armed with heavy chains called ticklers which flail at the seabed and drive bottom fish or shrimp up into the water column where they can be captured in the net. Target species include slow-growing rockfish which can take decades to reach maturity, cod, Pacific ocean perch, pollock and flatfish. In dragging, a practice common elsewhere although not used in BC waters, large chain bags which can weigh more than a tonne are pulled across the seabed to scoop scallops, mussels, oysters and sea urchins off the bottom and clams out of the mud. Typically, a boat will drag two such dredges at a time.

But in addition to the lucrative target species, the gear used in bottom trawling and dragging can be responsible for massive and indiscriminate by-catches of unwanted species which are mostly discarded. Studies have found as much as 40 percent of the trawl by-catch in BC is sometimes thrown back into the sea, dead or dying. Marine biologists say that bottom trawl gear disrupts and destroys the habitat for countless other creatures like seaworms, sponges, anemones, delicate tree corals, microscopic animals which burrow into the mud and non-commercial shellfish species which hold possibly important and often poorly understood roles in the complicated web of ocean life. Some of these are slow-growing creatures that can take centuries to fully develop. Sponges can live for fifty years, the west coast geoduck for more than 140 and the Atlantic quahog for more than 220. Whether the ecosystems these long-lived, slow-growing species occupy can recover from massive disruption, scientists point out, remains a matter of theory and optimism. Only our descendants will know the answer for sure—and that will be around the year 2200. All these complex, often fragile bottom structures, from ripples in sandy seabeds to corals to abandoned clam shells, provide shelter and protection to uncounted species of sea life. Destroying them is rending the very fabric of ocean ecosystems and with them, perhaps, the future foundations of once abundant inshore and deep-sea fisheries.

Although fish catches worldwide grew sharply through mid-century as industrial models and new technology transformed the killing capacity of commercial fleets, regional takes of fish peaked anywhere from seven to twenty-five years ago and are now in decline, often steep decline. Catches in the northwest Atlantic have fallen by more than 40 percent since they peaked in 1973; in the southeast Atlantic by more than 50 percent; in the east central Pacific by 31 percent; in the northwest Pacific by 10 percent; and in the northeast Pacific by 9 percent.

"Bottom trawling and other mobile fishing gear have effects on the seabed that resemble forest clearcutting, but affect an even larger area," warns Elliott Norse in a paper published in the *Journal of Conservation Biology*. "Trawling crushes, buries and exposes marine animals on the seabed, destroying habitat by plowing about half the world's continental shelf—roughly 150 times the forest area clearcut—each year." World wide that's an area, by way of comparison, larger than India and Brazil combined.

Marine biologist Don McAllister, president of the Ottawa-based Ocean Voice International, is gravely concerned over the fate of an extensive family of brightly coloured tree-shaped corals known as gorgonians or seafans. In some places these corals can live for 500 years and possibly more than 1,500 years. "Gorgonians are likely being clearcut far more rapidly than they would be displaced. But coastal management has not yet accorded these coral groves any particular protection," McAllister is quoted as saying in a recent article. These brittle, delicately-branched structures are most common in deep, offshore waters of the continental shelf. In BC, for example, they can form entire submarine forests which provide shelter for rockfish, sea stars, octopus and other life forms.

The only way to ensure the survival of the beautiful groves of seafan corals and the equally rich but largely invisible cosmos of the muddy bottom, says the World Wildlife Fund, is to set aside some areas where all bottom trawling is simply prohibited. The alternative is a leap into the environmental unknown. "The sea's equivalent of ancient forests are becoming cattle pastures by default," say Norse and his colleague Les Watling in a paper in the *Journal of Conservation Biology*. UBC's Daniel Pauly is equally succinct. "Though sometimes tempted by pessimism, I believe that we humans will, in the next Millennium, find ways to match our numbers and our demands with what our planet can provide (this is not so for the time being)," he wrote. "This will require that we abandon rape and pillage as our major mode of interaction with natural resources."

Marine Bombs and Other Phenomena

The storm that lashed Vancouver and the coast was what meteorologists call a marine bomb: a low that accelerates from irritating disturbance to lethal nightmare so fast that sailors can't even run for shelter. The yacht *Valentina*, caught in open water off Cape Flattery, was struck with such sudden violence that its sails were shredded and it was abandoned to seas that reached twelve metres at peak. Off Tofino, gusts pulverized the deckhouse windows of a big 30-metre commercial fish packer. These winds were so strong that they destroyed the wind meters at many of Environment Canada's more exposed automated monitoring sites, renewing the debate over plans to de-staff coastal lighthouses—and perhaps the debate deserves to be renewed.

Marine bombs are not uncommon on this coast. On their fringes, winds routinely reach hurricane force, capable of exerting pressures of more than 400 kilograms per square metre, enough to tear off roofs and flatten weak buildings at impact.

During this particular storm, off the sparsely populated north end of Vancouver Island, hurricane-force winds held steady at 165 kilometres per hour and gusts were recorded at 200 kilometres per hour. At Trial Island, off Victoria's pricey residential waterfront, winds blew steadily at 93 and gusted to 111 kilometres per hour, severe enough to uproot and snap off mature trees and even power poles. Gusts of near hurricane force were recorded in Haro Strait and at Saturna Island, just 60 kilometres from downtown Vancouver. But the storm that slashed across the south coast, leaving commuter chaos, blackouts and shipwrecks in its wake, was a commonplace event for the Northwest Coast, the first of the one hundred or more winter storms that can be expected between October and April. This one just happened to track through a heavily populated part of the coast, intruding briefly into our urban comforts and requiring city-dwellers to take more notice of the natural world than we usually do.

"For many years I was self-appointed inspector of snowstorms and rainstorms, and did my duty faithfully, though I never received one cent for it," noted the great American naturalist Henry Thoreau in his journal on February 22, 1845.

Here in the Northwest Coast our storms are matters of billions—not in cost, although some of the fiercest are damaging, but in the wealth they generate. Most of us tend to focus on the dollar value of damage—insurance settlements and the rescue operations that accompany an extreme atmospheric event—or on the tax dollars spent each year maintaining space and ground-based sensors that gather the information required by thousands of scientists employed to predict storm tracks and provide accurate localized weather forecasts.

Yet these maritime disturbances, which arrive at the coast with the regularity of loaded freight trains, are also the great atmospheric engines on which our whole provincial economy is based. They pay for our schools and hospitals and highways and municipal elections, although politicians faced with traffic snarls and downed hydro wires curse them more often than they give thanks. Winter storms are what fill the rivers that generate hundreds of millions in hydro-electric export dollars each year. They replenish the reservoirs that provide the cheap power that drives and gives competitive advantage to our manufacturing industries. British Columbia has almost 10 percent of the entire world's fresh water supply because of these storm systems. And without their ability to soak the ground so thoroughly, we'd have few of the fast-growing forests that grace the province with fully half of Canada's softwood inventory. Our winter storms can be said to create 300,000 direct jobs for the economy. Such systems set the annual snowpack, which in turn provides a winter diversion for the wealthy elites at resorts like Whistler. But, more importantly, that snowpack also provides the gradual release of spring and summer melt. It cools our rivers and carries a billion dollars worth of baby salmon to the sea on each spawning cycle. All these resources which derive from the great storm cycles of the north Pacific have one thing in common: the potential to bless British Columbians with an endlessly renewable foundation to economic life. They also have values that cannot be entered into the accountants' ledgers.

The psychological esthetic they create helps define us as a distinct maritime people. Artists from Emily Carr to Toni Onley capture an inventory of the lush greens and pale greys, the vivid contrasts and subtle mergings that characterize the interplay of west coast light and landscape. If bold sightlines and ice in the blood are what define our prairie cousins to the east, fog, spindrift and salt spray, the coming of October rains and April avalanches are what link BC's culture to the sea. However much the acolytes of plastic and silicon insist our only future is wired to the neural net of fibre-optics and global data flux, west coast storm-watching remains a popular rite so long after Thoreau took note of the phenomenon. At Pacific Rim National Park, where the fronts sweep unimpeded from the Ryuku Trench and the Aleutian Abyssal Plain, a new administrative routine emerges. The mesmerized must be warned that every storm surge has the power to reach up and pluck the careless from the beach.

For all its violence, the marine bomb begins life as a gentle feather of air over glossy swells southeast of Japan. A faint wisp of it curls off the southern edge of a convergence of two atmospheric jet streams, one carrying cold dry air out of Siberia, the other warm, moisture-laden air from the tropics. Both travel eastward across the Pacific. At the point of convergence, passing through the atmosphere at about 5,500 metres of altitude, the stream wobbles, buckles and a small area of low pressure protrudes from its edge, drooping southward like an invisible nose. This is the beginning of the process called cyclogenesis. Around this centre, driven partly by the spin of the planet, partly by the temperature differential between colder continental air and the latent summer heat still trapped in the oceans, the air begins to stir in what's known as an extra-tropical cyclone. In simplest terms, storms like this are born of the relative differences between low and high pressure centres. Then life is breathed into them by the differences in temperature between land and sea, which cool and heat at different rates. Physicists would recognize this process to be a textbook example of Newton's Second Law of motion—so elegantly described by a "pure" thinker without access to any of the sophisticated measuring devices that characterize our own science—which governs velocity, acceleration and direction proportional to the force applied, in this case, the movement of air from high to low pressure points in the atmosphere and the energy subsequently released by the process.

Pat Morin of Environment Canada's marine weather unit in Vancouver first noticed the tiny mid-Pacific anomaly in satellite images. So did Jamie MacDuff in the federal agency's computer-filled weather forecasting office at the Victoria International Airport. "It started as an innocuous little low pressure point out there in the mid-Pacific," says MacDuff, "but all the lifting dynamics were there. We had our eye on that thing the whole way."

In an era of deep Coast Guard budget cuts, meticulous marine weather forecasts assume even greater importance, which is why there is the rising political clamour over decisions to downsize the human observer component previously provided by lighthouse keepers. Nobody wants a replay of 1975, when a similarly fast-moving storm sank twelve commercial vessels and drowned thirteen sailors. Or the Black October of 1984, when another marine bomb capsized eight fishing boats, sank six and drowned five crew.

So, for all its apparently innocent beginnings, Environment Canada forecasters pored over the satellite images, the automated reports from robot monitors and the sophisticated numeric models spewed out by American supercomputers. At first, the distant perturbation wasn't dramatic, just a few black lines on a map—isobars: the joining of places of equal pressure that meteorologists employ to better conceptualize in three dimensions the invisible patterns taking shape in the air. When the weather system was crossing the International Dateline, the isobars were widely spaced, barely discernible as a distinct shape. These transient cyclonic phenomena are the most abundant of the large atmospheric mechanisms which determine the weather patterns dominating BC's west coast marine climate.

October and November are the dirty weather months on the West Coast. From then until spring, there occur from ten to fifteen major blows per month, although not all of them are felt in urban Vancouver or Victoria. But some are created in conditions that fashion them into a major marine bomb. Perhaps these storms seem more dramatic near the autumn equinox because we are reluctant to let go of the normally tranquil summer weather. But these equinoctial storms do tend to be more violent, in part because, as the northern latitudes tilt farther from the sun and less radiant heat reaches the earth, the continents begin to cool faster than the oceans. Thus, the differentials in thermal gradients which drive the storm cycles become most pronounced. And, according to Newton's laws, the sharper the differences, the greater the possibility of extreme weather.

Out in the Pacific, thousands of miles from land, the warm stream of air from the tropics began to suck up moisture from the ocean below. Then, as warm air does, it rose, spiralling upward and counter-clockwise around the point of lowest pressure. As it encountered the colder, drier air from Siberia overhead, it shed moisture as heavy rain. As it dumped its moisture, the developing system released latent heat back into the atmosphere. The warming triggered more evaporation, more precipitation and the air began to circulate faster. With the earth spinning beneath it and the jet stream above travelling the opposite direction, the low pressure centre began to track eastward across the Pacific. As it went, spilling cold air to the south, it created its own cold front, propagating winds of 8 to 15 metres per second. At this point, while the low pressure system was intensifying somewhere over the submerged mountain range called the Musicians on the undersea charts, the first of Environment Canada's weather warnings were issued in Victoria and Vancouver. As the system moved east with the accompanying cold front to the southeast promising to channel the most intense winds across the south coast, the isobars on the satellite maps had begun to narrow. Wind and heavy rain advisories were issued. The storm began to pick up momentum as it moved toward the coast. In twelve hours it moved more than 500 kilometres. The following morning the isobars had tightened sharply and the low pressure centre was deepening explosively. By noon the isobars on the computer-generated maps had condensed into a dense, black vortex of numbers. Prognosis models chattering from printers linked to computers in the US weather agencies showed a marine bomb in the making.

The parameters used to define a marine bomb are both strict and narrow. It is a transient low pressure centre in which the barometric pressure drops more than 24 millibars in 24 hours. "This one dropped 43 millibars in 24 hours," says Morin. "It was a good one. Not the worst in depth and intensity, but I'd say in the top 5 percent."

Fifty years ago, moving toward the coast at almost 50 kilometres an hour, a storm of this magnitude would have caught all but the most weatherwise by surprise. Experienced sailors might have spotted the pale wisps of cirrus in the high atmosphere that precede a storm front by as much as 300 kilometres. The freshening breeze, the barometer falling by one millibar per hour and spatters of wind-borne rain from denser, lower clouds called altostratus would offer other signals. But within hours the wind would have freshened to gales. The barometer would be falling by two millibars an hour and the arrival of denser stratocumulus along the horizon would give two hours notice of a full storm front. Then the sea would abruptly vanish under foam and spindrift, the waves would pile up to the size of a two-storey house and the skipper would pray he was far enough out to avoid the lee shore. BC's storm coast is called the Graveyard of the Pacific for a reason. It has claimed more than two hundred ships this way.

Today, with computerized radar and satellite imaging, such a storm is often a one-day wonder for media only because, for all our jokes about the unreliability of the weatherman, the forecasts proved precise and accurate and permitted us to evade disaster.

The front accompanying the storm centre crossed the Lower Mainland exactly as predicted, lashed the coast with high winds and heavy rain, as forecast, and then passed inland to die. There, starved of the moisture uptake and discharge that powers every cyclone's atmospheric engine, the storm dumped its final load of precipitation as snow on the Hope-Princeton highway and broke up.

And, certain as the sun will rise tomorrow, every October a series of storm systems will begin forming somewhere over the dark and restless Pacific. Which of them might become the next marine bomb will be determined only by Isaac Newton's immutable laws. Equally certain is the fact that, difficult as we find these storms to live with, we couldn't long survive without them.

WEST OF TIN CAN CREEK

Salish Drive curls down from the affluent estates of Southwest Marine Drive to the modest bungalows of the Musqueam Indian Reserve just west of Tin Can Creek. At the band's longhouse, just beyond the framed skeletons of a new subdivision, the broad sweep of blacktop suddenly reverts to gravel, then regresses to a dirt track at the overgrown houses of Chinese truck gardeners abandoned half a century ago. The road finally peters out in a narrow, rutted tunnel of salmonberry canes, dense thickets of scrub willow and brambles that almost seem to erupt from the black alluvial soil. Shadows here take on the luminous quality of light passing through jade. Just beyond the tangled wall of undergrowth can be heard the muscular brown hiss of the Fraser. To come this way is to travel back in time, from the bustle and din of the future's glass and steel and vinyl siding to the ancient landscape that was here when Vancouver was simply the name of a midshipman serving at the pleasure of King George.

Few of the 1.8 million immigrants who have flooded into Vancouver over the last hundred years know much about the People of the Grass, yet the X'muzk'i'um—the Musqueam—have been on this same site since Moses led the Israelites out of Egypt. This wild, luxuriant fragment just south of the manicured greens of Shaughnessy Golf Course is Mali, once site of the biggest longhouse on the coast. From it T'semalano, the grandfather of Vincent Stogan, came forth to greet and intimidate Simon Fraser in 1808. On this sunny afternoon, the leaves still sparkling with drops of the previous day's rain, tiny wild canaries make vivid splashes of yellow where they bathe at the edge of mud puddles. The fragrance of wild Pacific roses mingles with the sharp tang of sea salt and saturates the heady air.

Wendy Grant-John—one-time poster girl, former actress, cannery worker, super mom, colour-bearer of the powerful Sparrow family, hard-headed businesswoman, three-term chief, scholar, grandmother, successful politician, stateswoman, famous personage—simply closes her eyes, leans back and breathes in the sweet scent. She inhales it as though it might purge her burdens. They have been mounting lately. Her campaign to replace Ovide Mercredi as grand chief of the Assembly of First Nations has been interrupted by funeral rites for a multiple drowning at Tachie, 600 kilometres north of Vancouver. Her husband, Ed John, is a tribal chief of the Carrier people there. And now she's back to check up on her widowed seventy-two-year-old mother Helen, still recovering from recent brain surgery to remove a dangerous blood clot. "I come down here when I want to come to my senses," Wendy says. "I make time to come here. I believe in the voice of our ancestors. I still think it's here. I came here all the time as a child. I never knew any fear here—well, only once, delivering the paper to a house where they had a big dog."

That was almost fifty years ago, but as she speaks, eyes still closed, still drinking in the scents and sounds, the tender girl emerges from the tough, world-wise woman. "It was a wonderful place to grow up. My grandmother was a longhouse woman here. I spent a lot of time here with my grandparents. You could see right down to the river. I spent hour upon hour running on the log booms out in the current. The men would pull up their canoes on the sandbar and we used to come down and swim. The eulachon used to come up here so thick you didn't even have to set nets." She was Wendy Sparrow then, the long-legged, carefree eldest daughter of Willard Sparrow and Helen Malcolm, the first of ten kids that would bless the union, part of a vigorous family that has provided the band with five elected chiefs, most recently her younger sister Gail.

The great river coils through her dreams of childhood as it courses through the lives of all the Musqueam people, just as it connects her to her husband's distant people. The first sockeye are in the Fraser and her sons and brothers and cousins are all out fishing. The fish in their nets are Early Stuarts, silver "runners-to-the-sea" going home to Tachie so far to the north. Her eldest son Trent began saving money for his boat before he was a teenager. He'd go to the golf course and scour the rough for lost balls, selling them back to golfers. "He got a 16-foot boat and went out fishing when he was fourteen," she says. "He's been on the water ever since. I'd just die watching them out there in that little boat. But he respects the water."

To discover the profound importance of fish and fishing to the Musqueam, one need only spend an hour in the band's Roman Catholic cemetery. Once the ancestors were interred in beautifully carved burial boxes, a testament to the wealth of the village. But these were removed to the National Museum in Ottawa in 1931. Today, the place still pulses with the old ways. Graves of elders are marked by gifts from the grieving: a well-used fielder's glove, a teenager's soccer trophy, a little girl's birthday balloon displaying a bright red heart. "Those people who have gone to the other side are very much part of you," she says. "You have to pay them that respect even though they're not there." The complicated genealogies of ancient families tangle like the skeins of a net in this culture, the old names and the new: Teeochlash and Whywhialack, Whunathit and Xullilum, Grants and Guerins, Sparrows and Stogans, Charleses and Campbells, all tracing descent from a great man who had two wives two hundred years ago. It's a mariner's graveyard. The black marble stones favoured these days are engraved with images of the sea. Kenneth Peters mending nets. Arnold Guerin's double-ender, moored for eternity on glassy waters. Richard Gordon Point's gill-netter, *Sailing Home.* And David Henry Joe offers a twenty-one-year-old fisherman's wry humour to the passerby: "God grant that I may live to fish until my dying day, and when my final cast is made, that in His mercy I be judged good enough to keep."

Willard Sparrow was a handsome young fisherman working out of Steveston at the end of the war when he met Helen, an innocent from New Westminster who says she didn't even know what an Indian reserve was. It was a remarkable event for a white woman to marry a Native Indian man in 1948 when aboriginals still didn't have the right to vote and women had almost no rights at all. "It was two years before I met his parents," Helen recalls. "He wouldn't bring me down here because he was ashamed of it. He said he lived up by Point Grey—I thought he meant up by the school. I was pretty naïve, I guess." But Wendy says this honesty speaks to her mother's great strength of character. Helen not only moved to the Musqueam reserve after her marriage, she stayed for the thirty years following her husband's premature death to become one of the community's most respected elders. "My mother ended up a widow at the age of thirty-eight with no life insurance and ten children to raise," she says. "I look at it now from the age of forty-eight—if I had been left to raise my four children alone with no support—I don't know how she did it."

It was hard, Helen agrees. There was the poverty—she often had to send her kids to school with nothing on their bread and white margarine but a sprinkle of sugar. And worse, there was the patronizing smugness of the Indian Affairs department bureaucrats.

The nurse who'd simply walk into her house unannounced to inspect it. "I got $8 in the mail once. I had no idea what it was for. It turned out it was for having the cleanest house during one of their spot checks," Helen remembers. But her mother's quiet strength is what Wendy credits with getting her through her own trials: the racism she endured throughout her school years; the loss of her father when she needed him most; her own bout with the alcoholism that killed him at the age of forty; the birth of a son who required heart surgery when he was only eleven days old. "If it wasn't for her, I'd be a statistic right now."

Wendy still winces when she talks about the trauma of school in the mainstream. "We were the first Indian kids there, the first ones not to go to residential school. It was hell. My first day at Southlands Elementary School I was told by another girl that I couldn't come in the same gate as the white kids because I was from the reserve. I remember it so clearly. I remember her name. I bet she doesn't remember saying it to me, but I sure do." Then there was the grade eight sleepover when the last-minute phone call came: there'd been a mistake—they'd invited one too many kids. But she knew the real reason she was being uninvited was because parents had discovered she lived on the reserve. "Once you hit Marine Drive, it was a different world," she says. "Honest to God, the way people talked about us: if you'd closed your eyes you'd have thought you were in Mississippi." At Point Grey Senior Secondary, she says, the boys were despicably cruel. "Sometimes they'd come right out and call me an f---ing squaw. You were just considered a dirty Indian from the reserve. To be told you aren't wanted like that when you're fifteen—it hurts. And it didn't get any better. I quit Point Grey six months before graduating from grade twelve. I wouldn't have lasted that long if it weren't for two women who were counsellors."

Willard Sparrow, however, saw the woman his daughter could become. Every summer until he died, he'd taken her fishing with him in the gillnetter he named the *Miss Wendy*. "Remember who you are, who your family is and where you come from," he told her. "These aren't empty words. Your responsibility is to help those people understand who they are and not only what they've done to us, but what they've done to themselves." Although he was a poor man, serving as the band's first administrator and working out of his bedroom, he sent Wendy uptown to the Blanche Macdonald Finishing School to obtain the polish he knew would serve her well on the other side of Marine Drive.

As a teenager, she developed a natural, serene beauty that has remained with her into maturity. It won her a couple of modelling jobs—one of them posing as a Mexican for a Canadian Pacific Airlines poster—some television parts and a film role opposite Len George in a bad movie. "Len George is a very fine actor," she says. "Wendy Sparrow was not." She tried waitressing at Muck-a-muck, the West End restaurant featuring smoked salmon and Indian ice cream for upscale tourists. She lasted one shift at that job. So she went back to the river and the line at the Queen Charlotte Cannery, washing fish and working the clinker putting lids on cans of salmon. Later she graduated to packing cases. She was never fast enough, she admits, to work on the cutting line. In winter, she worked the herring.

Her father, in the meantime, had built a room off the back of his house to serve as the band office and he was making things happen. He turned his own property back to the band and led a move to reassert collective title. Next he lined up the first Canada Mortgage and Housing Corporation financing for on-reserve mortgages and persuaded the band to develop its own housing. "All these homes are paid for by the band," Wendy points out. "People think our houses are paid for by the government. That's not true here." And then, suddenly, her father died.

"It was very, very devastating. When he left us, I became very dysfunctional. I was an alcoholic. I had a real problem." What saved her, she says, were three things: Her mother's steadfast belief she'd come through it; the realization that her new baby's heart condition wouldn't tolerate her conduct; and six women who got her involved in learning to weave traditional woollen blankets. Their names are Margaret Louis, Christa Point, Barb Cayou, Linda Joe, Robyn Sparrow and Leila Dan. "The experiences I had with those women—nobody could put a value on them. The sharing of life's difficulties and what was happening in their families made me see the difficulties and needs of our whole community. Working with those women on weaving, everything came together. They brought me out. They pushed me to go into politics." The rest, as they say, is history.

In 1986, at the age of thirty-seven, she was elected to the first of three consecutive terms as chief of the Musqueam, one of the first women to be elected to such a role among the 196 Indian bands in the province. Her tenure was marked by a series of dramatic political milestones. Under her leadership, the Musqueam won two landmark cases before the Supreme Court of Canada—Guerin and Sparrow—which clarified and defined the entrenchment of aboriginal rights in the Canadian constitution. At home, she guided the birth of one of the first effective aboriginal taxation systems in BC, launched the first aboriginal commercial fishery in Canada and persuaded the Vancouver School Board to include Musqueam cultural history in the curriculum. Today there is a First Nations progam at Point Grey Senior Secondary where she was once so mercilessly tormented. She served as president of Musqueam Holdings Inc.—which has as its subsidiary Celtic Shipyards—coordinated land management for the band and helped establish the treaty negotiation office. And she has contributed widely to the broader community, representing Canada on the Pacific Salmon Commission, advising the province on aquaculture and serving as a director of both the Royal British Columbia Museum and the Children's Educational Foundation.

Her proudest achievements, she says, are on her own reserve and some of them seem small—a playground, for example—but loom large in the eyes of parents. She helped set up a safe house so that kids at risk need not be removed from the community, established an elders drop-in centre which will eventually become residential, launched a fisheries commission and bought back the Fraser Arms Hotel, which is built upon one of the ancient Musqueam village sites. "I happen to believe that this is the greatest country in the world to live in. The society that I'm privileged to be part of has such strength and resilience—that's what we have that we should come together and celebrate. It's going to be tough, but it's time."

The Reef Where Time Began

A silken, lightless sea murmurs against the hull. The forest crowding the water's edge grows restless on an unseen wind. Beyond it, the duskier bulk of island against distant headland, all looming against a moonless sky that is as deep and dark as time itself. This is the lonely coast, the wild coast, equal parts an ancient, mist-draped past and British Columbia's unfolding future. Rainwashed, windswept, still far less populated now than it was a century ago, it sprawls northwest from the smoggy, bustling Lower Mainland in a convoluted, skewed topography of tilted plates, snow-clad massifs that soar four kilometres high, fiords that plunge into a sea as black as obsidian and a maze of jewel-like islands scattered upon the white-laced sea.

Offshore, arched like the surfacing spine of some sea creature from the Jurassic, flanked by the foam from its five-metre tides, is an archipelago of 1,884 islands that's been called the Galapagos of the North, a stunningly beautiful world of unique biological marvels. Here immense eagles soar on the thermals, the rarest and the tiniest of bats huddle in the warmth of geothermal vents, the world's largest black bears pad the beaches unmolested and some of the most fascinating forests are not on land but under water. These kelp forests, waving in the nutrient-rich currents, are home to an astonishing variety of life—bat-shaped starfish, carnivorous sea anemones, giant eels and huge octopus. Enormous jellyfish pulse like moons in the water. On the bottom, a startling palette of rockfish—the Vermilion, the Blue, the Copper, the Yelloweye, the Chilipepper, the Canary, the Painted Greenling. Past them cruise the swift sei and minke whales, lazier humpbacks, deep-diving sperm whales, orcas and grays. Among the fronds slide seals and sea lions the size of a small car—so clumsy on land, but in their natural element lithe and graceful as Olympic gymnasts.

I'm on the southeast shore of Moresby Island, 720 kilometres northwest of Vancouver and 200 kilometres off the mainland, to look over the proposed marine protection area that conservationists want Ottawa to extend 10 kilometres into the sea surrounding Gwaii Haanas National Park Reserve. Years ago, the federal and provincial governments made a commitment to the Haida that the adjacent sea bottom and the water column above it would receive protection similar to that given Gwaii Haanas. Some encouraging steps have been made in that direction, says Sabine Jessen of the Canadian Parks and Wilderness Society, but what's needed is committed action. "The fantastic marine biology here, the incredible diversity of species demands that we begin to manage on an ecosystem basis rather than the basis of individual species," says Jessen. "Total ecosystem management is what it's all about now." The Haida, too, are increasingly concerned. Gujaaw, descended from the Raven clan of Skedans, co-chairs the Gwaii Haanas management council. He says the traditional species-by-species approach to fisheries management is worrying. Tensions are already running high over commercial in-shore herring fisheries and commercial sport fisheries for salmon. "The pattern has been fishing to depletion," Gujaaw says. "We know that from whenever the urchin boats and geoduck boats go through. In the last twenty years we've watched abalone wiped out, rockfish is being wiped out, just about everything you can jig for is being wiped out by commercial fishermen," he says. "To me, the cultural reasons for protecting the land are ultimately connected to what goes on in the sea."

Perhaps with fish stocks collapsing world wide, a planetary-scale die-off in coral reefs, the cod disaster on the east coast, BC's own wake-up call on salmon survival and marine scientists everywhere calling for marine sanctuaries where breeding stock can find refuge, the message is just beginning to sink in. "What goes on in the water has direct impact on what goes on in the terrestrial park," concurs parks scientist Doug Burles. "The greatest extraction is from the water. The herring fishery has a great impact on the land because salmon, which rely on herring, are major contributors to the nutrient cycle in its ecology. A lot of the human activities here are on the water. Having more control over that logically means we can have more control over the ecosystem." So Jessen's lobbying—with bureaucrats, with the Haida and with non-aboriginal interests—takes on a fierce intensity as she argues that communities must sustain the environment if they expect it to sustain them, and that a marine protection area will give local people more say over how the immediate environment should be managed.

The idea of large-scale underwater parks is part of a growing worldwide movement, rooted more in science than emotions. It sees an intimate connection between what happens on land and what happens in the contiguous ocean. This vision argues for the protection of both terrestrial and adjacent submarine landscapes as a way of saving complete ecosystems that extend from mountain peaks to ocean deeps. "We have to get the idea across that there is a marine environment in Canada that warrants protection as well," says Steve Langdon, the top federal bureaucrat in Haida Gwaii who's responsible for the national park reserve. "The biggest challenge is getting acceptance for the marine conservation area concept—getting the stakeholders to buy in. Stakeholder buy-in, community buy-in, those are essential components of our strategy," he says. "In Gwaii Haanas in particular it's pretty difficult to separate the marine ecosystems from the terrestrial ecosystems." Salmon and herring, eagles and ravens, bats and plate tectonics, bears and sea lions, porpoises and killer whales, ancient cedars and giant kelp, people and plankton—all are bound together in the timeless cycles of energy exchange that govern their environment.

Nowhere are these inter-connected cycles quite so rich or intact as in Haida Gwaii. Cold, nutrient-rich water from the ocean trenches sustains life in the shallows. Herring spawn and feed there. Salmon feed on the herring then return to the rivers to spawn. Bears and eagles glut on salmon. Eagles scatter rotting carcasses and gorging bears defecate in the woods beyond the river banks. Both distribute vital nutrients beyond the immediate riparian zone. The understorey at the river's edge, with benefit from some of the highest value fertilizer available, then erupts with growth, providing shelter and nutrients for the next generation of salmon in the cycle. And people, too, have been turning on this wheel of life since time immemorial.

I've risen early while the others sleep because I want to see the first light fall on Xa'gi, the place where the Haida say time began, at least as humans measure it, and the world as we know it came into being. The name means "striped" and is said to refer to alternating intensities of light and darkness, both in the rock and in the changing texture of the seascape. But it might also refer to the alternating levels of existence that comprise Haida cosmology, a ten-layered universe ranging from the world of stones to the world of unseen, supernatural beings, all tied together by one cosmic cedar which is the unifying metaphor at the heart of Haida Gwaii. This immense, mythical tree, its roots tapping the currents of the underworld, its trunk supporting our world, its branches gathering the occult power of the firmament, connects everything and redistributes cosmic energy to keep the universe in balance and all beings in harmony. Like water passing through the hydrologic cycle, like the transformative chemistry of salmon changing from shimmering horde to bear scat to riparian foliage to the baseline nutrients that will sustain a new generation of salmon, people themselves travel between these cosmic levels, now in the present world, now dying to experience a new incarnation in the realm of the supernatural, now returning to this world again, reborn into the infants of the next generation.

At Xa'gi, everything is reduced to the yin and yang of elementals. Wind. Water. Rock. Light. By night there are no artificial lights on this horizon, no lights on any shore, not even the distant habitation glow to which our urban lives make us accustomed even in small communities. There are only these layers of black and blacker-than-black, and above us the stars, a river of them suspended in the heavens, glittering points so bright they almost pierce the eyes. Behind them lies the softer glow of the Milky Way. What the eye perceives is a band of glimmering messages from the distant edge of the galaxy, light waves dispatched before the beginning of our world, a reminder of Haida cosmology—their vision of a universe of cycles within cycles, mysteries within mysteries.

Behind me, in the cabin, the sleepers stir and mumble in their berths, dreaming about what? The boat swings and shudders on the current. The turning tide makes night music under the stern, small eddies of phosphorescence pinwheel away and vanish again with a faint hiss. Out there something big swirls on the surface, slurps and sighs, leaves a widening dimple. Seal? Sea lion? Porpoise? My brother told me once of watching a huge shark rise silently from these depths under the stern of his fishboat, open its serrated mouth to the galley garbage like a scene from the movie *Jaws*, then sink back into the inky sea. Another skipper on the docks at Queen Charlotte City tells of a Great White out in Hecate Strait mistaking the blue-painted bottom of his prawn boat for the soft underbelly of a whale. The slash marks from its teeth ran 12 metres from stem to stern. Yet another tells of hearing a splashing off his bow during a midnight drift and finding a huge ball of pygmy sperm whales cavorting on the surface under a full August moon, "the sperm moon."

For all our science, there are more mysteries in this sea than we begin to fathom. Waiting for daybreak, the air is soft for all its sharp undertang of salt and cedar. Why is it that at night, senses which seem to sleep during the day suddenly come alive, allow us to pick out the faint scent of earth wafting over water, the odour of decaying needles, punky wood and fungus, the pungent iodine smell of drying seaweed? At night we hear things we never hear by daylight: the rustle of a bird folding its wings, the lap of wavelets, the distant squeal of dry branches rubbing in the wind, the underlying thud of our own hearts.

Dawn in these northern latitudes is not the thunderous affair that it is farther south. There's a faint lessening of the darkness to the east, then a splinter of light over Deluge Point, named in 1878 when torrential rains drove George Dawson's geological survey team back to huddle "soaked, chilled and disgusted" in their schooner's cabin. Finally, Xa'gi brightens against the waning darkness. A silver line of surf off the end of Bolkus Island defines the precise point at which sea and sky meet. Here, where the sea breaks over the reef, Haida legend says the first land emerged from the primordial sea. This still feels like a place of magic and power. Anthropologists describe such reefs as representing in Haida myth a doorway between planes of being, a transition zone between cosmic dimensions that the old ones believed to be defined by the tides, of which there are many—tides in the sea, tides of light, tides in the seasons, tides in life itself.

Here Foam Woman made her appearance, riding the reef up out of a great flood. Tangled in her hair was the branch which would bring life to a barren landscape. Around Xa'gi gathered other supernatural beings, but whenever one attempted to come ashore, Foam Woman drove it off with lightning bolts that flashed from her eyes, earning her also the name She-of-the-Powerful-Face. From this titanic, elemental figure are descended all the Haida lineages of Raven: the figure of tricks and transformation who dominates the cosmologies of British Columbia's coastal peoples, as Coyote dominates those of the inland cultures to the east. Foam Woman's descendants became known as the Striped-town-people and their descendants in the Raven clan spread to populate the area Europeans called the Queen Charlotte Islands, but which in their own language continues to be known as Haida Gwaii—land of wonders. And yet, perhaps the transition zone described in Foam Woman's story is also, as anthropologist Nicholas Gessler remarks, symbolic of the first contact between ancient mariners and a new land, newly risen from the sea. Archaeology lends some credence to the story. In the intertidal zones around Skincuttle Inlet is found the oldest evidence of human habitation, dating back perhaps 5,000 years, older than Moses and the Pyramids. Haida stories suggest the fusion of two distinctive cultures about 2,000 years ago, the first consisting of the original islanders of Foam Woman's Raven clan, the second consisting of newcomers from the mainland who formed the present Eagle clan.

Like those of all the peoples indigenous to the north coast, the myths and legends of the Haida are filled with floods and inundations. Some of them are almost certainly the metaphorical renderings of a collective memory from the dim outer fringes of human recollection, a memory from the last ice age and the sudden, violent fluctuations in sea level that accompanied the great melting of glaciers 12,000 years ago: the rupture of inland ice dams and the rebound of the earth in the absence of a burden that was two kilometres thick. Even in geology, the linkages between land and sea abound. It's thought that when the glaciers ruled and sea levels were much lower than today, these islands were separated from the mainland only by a narrow gut. Perhaps it froze and formed an ice bridge by which caribou, bears and people arrived. Perhaps the people came by sea.

Old Haida stories tell of a third, drowned island that was halfway to the mainland. Much of the shelving stretch of Hecate Strait between Graham Island and the mainland is shallow, a good deal of it 100 metres or less in depth. Oceanographic mapping off Cape Ball indicates an ancient extension of the islands which has been submerged for more than 10,000 years. Recent finds of stone tools recovered from a bottom trawl lend credence to the idea of an ancient Haida Gwaii now sunken beneath the sea like the Greeks' mythical Atlantis. On the other hand, the find may only be the chance recovery of equipment lost in some maritime disaster or cast into the sea as votive offerings in some unknown ritual. Still, the former idea is plausible, even appealing. Other archaeological sites—there are more than five hundred identified—indicate how radically the coastline has changed. Palaeobotanists know the landscape then was an open, shrubby, tundra-like expanse already giving way to parkland as the climate warmed, ideal country for large herbivores and the people who hunted them—until rising sea levels flooded them out and eventually isolated the islands from the mainland. "The old stories tell about the days when people took shelter in the grasses," Gujaaw says. "It seems odd to me to think of our people being here before there were even trees."

The floods of myth and legend may also refer to the region's propensity for submarine earthquakes and attending tsunamis. One mythological chief who destroys his supernatural enemies by inundation is named Great-Splashing-of-Waves and one of the surviving spirits is Always-Looking-Out-to-Sea. Interestingly, this particular incident is the one which marks the end of the age of spirit beings and the beginning of the time of humans. In many ways, echoes of this mythological interpretation resonate through what scientific reason leads us to understand of the geological origins of the islands that became Haida Gwaii. Although they consist of layers of both sedimentary and volcanic rock, it's now thought that the fragment of the earth's mantle on which they ride originated more than 200 million years ago in the South Pacific, somewhere off the coast of Peru. Sliding northward along the edge of the continent, finally colliding with it, twisted and tilted upward by unimaginable geological forces, rotated while passing over a thin place in the mantle known as the Anaheim hot spot—its track eastward under the continent can be traced in a string of volcanic cones which begin in the Masset plateau and pass through the Coast Range and on into the BC Interior—the islands took the form they have today. They are the nexus in a pattern of fracture and suture lines, the convergence zone where one continent-sized plate is driven beneath the other. So Haida Gwaii rides up like a stone canoe over one of the world's most geologically active subduction faults. The tension generated is released in frequent earthquakes—one of them, in 1949, registering 8.1 on the Richter scale. We know from records in Japan and physical evidence ranging from Oregon to Tofino that a tidal wave big enough to transport entire beaches inland struck this coast at the beginning of the eighteenth century, probably triggered by a great mega-thrust earthquake in the subduction fault. It's this complicated buckling and twisting where the North American and Pacific plates collide that give these islands their most unique characteristics. For one thing, the tectonic history has important implications for the generation and entrapment of large pools of petroleum. For another, the rocks are unusually rich in metallic minerals, particularly iron, copper and gold.

So a decision to set aside a marine conservation zone contiguous to Gwaii Haanas is not without a price. In 1988 when a series of fierce battles between loggers, environmentalists and aboriginal communities was ended with the federal-provincial agreement to establish a national park reserve on Moresby Island to be co-managed by Ottawa and the Haida, five oil companies still held drilling rights. They recently agreed to relinquish those rights in the proposed marine protection zone, although they retain them in offshore areas nearby. With a downturn in the economy and the other primary resource extraction industries of forestry, fishing and mining in trouble province-wide, offshore drilling is seen in some quarters as a panacea that would kick-start a new economic boom on the north coast.

Pressure to lift a moratorium on petroleum exploration in other waters surrounding the part of Haida Gwaii that's not protected is growing on the heels of a Geological Survey of Canada report suggesting as much as $200 billion worth of oil and natural gas might be found—and that promises to renew tensions on the environmental front. More than the geology of these remarkable islands is at stake. They are among the richest places on the planet for biodiversity. Many of the plant and animal species here are found nowhere else. Scientists speculate that some species were isolated in a "refugium" during the last ice age and left to evolve outside the mainstream, hence the nickname "Galapagos of the North."

Even the islands' biotic zones are strikingly different from one another. To the east, the islands descend from the rugged 1,200-metre-high San Christoval Range in a series of relatively gentle benches. These shelves continue their graduated descent under the sea toward the mainland. The mountain wall shields both the eastern slope and a scattering of islands along a ragged coastline notched with many bays, inlets and inside passages which provide protection from most of the prevailing weather. Safe anchorages, sheltered beaches and lazy river estuaries make the whole area a paradise for kayakers, sailors and recreational boaters. To the west, however, the islands present the North Pacific with an impenetrable wall of sheer cliffs. For sailors, this wild coast represents a hazardous step back into a largely forgotten past—even the most modern charts for this shore still carry warnings of "uncharted waters." It's not hard to imagine why: there are only a few keyhole inlets in which a mariner might hide from storms which routinely reach hurricane force and generate waves up to 25 metres high.

Just offshore, the bottom falls away into a 2,500-metre-deep abyss. Even here, the sea reminds us of the intricate connections of things: this water was in the North Atlantic—about five hundred years ago. Carried up from this cold trench on a conveyor belt of ocean currents, carried around to the sheltered waters of Hecate Strait, seasonal upwellings of nutrients sustain a rich abundance of marine life in waters so clear that visibility has been measured at more than 45 metres. Fish shoal around submarine pinnacles that rise to within a few metres of the surface, attracting other predators. The vast kelp forests sway in sheltered coves teeming with exotic Sailfin and Staghorn sculpins, Cabezons and sea stars, molluscs and nudibranches, crabs and sea cucumbers, red turban snails, sea urchins in beds so thick they number three hundred to the square metre, bat stars in an abundance that biologists describe as "staggering" and scores of other invertebrates. Sandlance, tubesnouts and herring spawn here in vast numbers; salmon follow this feed before pouring into fifty-six rivers to spawn themselves. Just offshore, ten species of cetacean cruise with the pinipeds—the seals and sea lions. Jaegers, petrels, gulls and a host of diving birds bob along the currents and tiderips. From the air, approaching from the south, the archipelago of Haida Gwaii rises out of the Pacific in a blue-green wedge shaped like an isosceles triangle. It might be a continuation of the spine of mountains that form Vancouver Island except that it's rotated sharply to the northeast and separated from Cape Scott by 200 kilometres of open ocean.

There are 138 islands in the group that comprise Gwaii Haanas, extending from storm-swept Cape St. James, scarcely 100 metres wide and just off the southernmost tip of Kunghit Island, to Tasu Sound about 115 kilometres north. All told, the terrestrial park covers 145,000 hectares. The marine protected area under consideration would add another 305,000 hectares. It is still, strictly speaking, a deep wilderness area. There are only two warden stations in Gwaii Haanas, although the Haida maintain four base-camps at culturally important village sites. (It's not unknown for tourists to attempt to make off with entire totem poles or other artifacts.)

To get to this remote shore, we dispatched our boat, a fast 28-foot welded aluminum hull rigged for diving charters, two days ahead of time. This enabled the skipper to carry an additional 300 kilos of fuel in jerry cans, expanding our cruising range in an area where there are no services, no refuelling depots, no immediate help in case of trouble. And we would need it: before we were done we'd have covered more than 500 kilometres of coastline. We took a float plane to Rose Harbour, a former whaling station that is now in private hands, a thumbnail of property on the beach that's surrounded by park. Whale teeth, many of them from baby sperm whales, wash out of the beach gravel by the bucketful, a grim reminder of the industrial-scale slaughter that once took place here.

Behind the cluster of cabins, fifteen-year-resident Goetz Hanisch, who says he arrived with "a guitar, a packsack and a sack of potatoes," walked us back through the thousand-year-old cedars to look at a long, silver shape in the moss. It was a half-finished 12-metre canoe, the adze marks still evident. What caused the builder to abandon it? Was he taken, perhaps, in the smallpox epidemic that turned most of the Haida villages into ghost camps that summer of 1862? With this sobering thought, we pushed on for Skun'gwaii, Red-cod-town: the World Heritage Site on Anthony Island, occupied for 2,000 years and the apex of Haida civilization until the destructive forces unleashed by European contact reduced its population by 90 percentalmost overnight. The shell-shocked survivors abandoned it in the 1880s.

Secluded in a cluster of rocky, Bonsai-like islets that shelter it from the violent storms marching in from the Gulf of Alaska, this is the most famous Haida village of them all. Surrounded on all sides by the wild ocean, so remote and isolated that even the language spoken here had begun to diverge from the mainstream, its people fearless sea-farers and astonishing artists, Nunsting was once home to twenty longhouses and a forest of amazing totem poles. On November 27, 1981, it was designated a World Heritage Site by UNESCO, important enough to transcend the politics of ethnicity and nationalism. If, as many art historians believe, Haida carving represented the apex of west coast artistic achievement, then the dazzling array of house fronts, memorial and mortuary poles of Nunsting represent the apex of the apex. The art here is something held in trust for all humanity. And if aboriginal art lay outside the mandate of a story about marine protected areas, who could resist visiting a site of such spiritual and aesthetic power? And yet, why not? The Haida of Gwaii Haanas were connected to the sea as surely as the great sea mammals they hunted.

We came ashore on a side cove to avoid disturbing the keel runs for those great sea-going canoes of two hundred years ago that are still visible on the main beach. Then we walked with Captain Gold, the designated Haida watchman, along a path lined with clamshells. He led us into the village site in a tawny beachfront meadow loud with bees, itself a lovely image of dappled sunshade, golden grasses just about to scatter seed, the deep green of salal and the gold-tinged green of mosses contrasting with the weathered silver of poles spectacular enough to stop an art lover's breath. "This is a living graveyard and you have to approach it with that respect," Captain Gold warned. "In 1862 there were over 300 people in this village and almost overnight there were 30."

The story of how the smallpox epidemic was deliberately brought to this most remote village remains deeply disturbing even now, 136 years later. Francis Poole, then a miner working in Skincuttle Inlet, revealed it a decade later in his *Narrative of Adventures and Discovery in the North Pacific.* When a European passenger fell ill with the virulent disease during a passage to Victoria in 1863, the ship's captain insisted on putting him ashore at Nunsting despite Poole's protests to the contrary. It's hardly possible that the captain was unaware of the epidemic which had annihilated most coastal aboriginal communities over the previous year. "Scarce had the sick man landed when the Indians again caught it. In a very short space of time some of our best friends of the Ninstence or Cape St. James tribe...had disappeared forever."

Soon the survivors abandoned the site. It lingered for a while as a seasonal camp. Today, there are only decaying remnants, a melancholy legacy of what might have been. Great beams sag into house depression pits, belying the grandeur of longhouses with names like "People Think of The House Even When They Sleep Because the Master Feeds Everyone Who Calls," "Thunder Rolls Upon It House" and "People Wish To Be There House." Poles have been defaced by the rot of time and in some cases by vandals, Captain Gold says: a museum curator who cut a face out of a pole with a chainsaw, sea cadets who climbed one and broke off pieces, passing fishermen who burned some. The most important poles to the anthropologists and art historians are already gone, removed at least to the provincial museum in Victoria for preservation almost half a century ago. That decision, too, is not without controversy. Perhaps these poles are supposed to subside back into the earth from which they sprang; perhaps that acknowledgement of the power of time is part of what they mean. "A lot of us say when we see that film [of the removal] that Raven is crying because they are taking away the pole," Gold says. "You know, just when the pole touched the ground, the heavens opened, thunder came and it poured with rain."

Yet even what remains at Nunsting has a curious power to move the most cynical observer. Handmade square-headed nails show where a copper was attached to a wealthy chief's pole. Slender bones protrude from a collapsing mortuary box that contains some long-forgotten parent's beloved child. The asymmetry of a memorial pole suggests either an established carver whose powers were in decline, or a young carver trying to learn his craft in a post-holocaust era with no one to teach him. Our guide stops at the pole of a house whose name is lost in time, lays his hands on a small, easy-to-overlook face carved into the side of a fierce bear. "This is the ghost that lives in the fur of the sea grizzly. He collects the souls of people who drown at sea. But this is a Tsimshian crest. How does it come to be here? Acquired by war? By marriage? By a potlatch gift?" The question, of course, has no answer. It speaks as eloquently as anything can to this coast's greatest tragedy: the abrupt and irredeemable rupturing of a rich ancestral memory from one generation to the next.

From the world of myth and legend, we next ventured out to the exposed west coast to see a colony of several thousand Steller sea lions, bellowing bulls of a tonne or more presiding on spray-blown rocks over huge harems of females barely a third their size, the surrounding swells seething with teenagers looking for action. A gang of fifty followed us for an hour, daring one another to see which show-off could dash closest to the boat. One after another, accompanied by wild excitement among those hanging back, the most daring would pop out of the water an oar's length from the rail, then race away. At Flatrock Island we looked for the rare horned puffin, ancient murrelets and Cassin's auklets. This naked block that erupts vertically from the sea like some huge Egyptian monument is a critical nesting area. In fact, tens of thousands of pelagic and shorebirds can be seen throughout Gwaii Haanas. It's thought that close to 750,000 seabirds—15 to 25 percent of BC's entire bird population—masses here. Orange-billed puffins return to hatch chicks after their winter foraging on the open sea, red-legged oystercatchers scurry along the reefs, diving birds are everywhere.

Then, accompanied by a contingent of Pacific white-sided

dolphins that whizzed in dizzying circles around the boat, we turned for the gentler waters of the inside shore and a mooring for the night at Skincuttle Inlet.

I watched the quicksilver morning spill over the place of myth where human beings first entered the world, and then we went to examine some places whence they had departed. On Jedway Bay, we found the remnants of an abalone station. Before the Japanese came, this site was occupied in 1791 by Chief Ucah, the most powerful leader in the region. The Japanese, never licensed and serving a secretive domestic black market, abandoned their operation shortly before the First World War. Now it sinks into the mossy ground like a foundering ship. Only a few bits of rusting equipment and some rotting wood mark the site, although the beach is a collector's treasure trove, white with the elegant tide-polished shards from the fine china tea services of long-forgotten women. When they left, for whatever reason, they appear to have left in a hurry. In the trees on a small point is the weathered headboard of Mrs. Taniyo Isozaki, only 30 when she died on March 21, 1913. An epitaph hand-lettered in blue marine paint peels away from the layer of white painted over the layer of yellow: "Sayonara since it must be, but the word is hard to say." Yet this tranquil grave and its poignant goodbye, so far from the world, has been recently cleaned up: a new cedar fence erected, some beautiful pieces of broken teacups laid in a pattern among newly planted snowdrops, Leopard bane and primula. Nobody we ask knows for sure, but somebody heard from somebody else that a Japanese man chartered a bushplane, stayed in-country several days, then pulled out leaving all his expensive camping equipment behind. Others say a commercial fishboat put in and the crew paid its respects by tidying up her grave. Just across the inlet, on a river estuary emptying into George Bay, another melancholy enigma of human departure: past the mud flat, beyond a sunny meadow blazing with buttercups, we find the time-blackened stumps of ancient salmon traps exposed by winter rains. They are as old, perhaps, as the Sphinx or the walls of Troy, and the traps are a vast complex, extending far up the river and into its tributaries.

Skincuttle Inlet, we later learned, once supported ten Haida towns, a busy hub of clan and family connections that spread across the whole archipelago. Today it's the home of ravens and eagles. But just north, where Burnaby Island pinches in against Moresby, comes a reminder of the bountiful fecundity of life. At Dolomite Narrows where the tide runs like a green river four times a day but the channel is completely protected from wind and wave action, there is more life per square metre than anywhere else on the planet. One survey collected 14,962 different animals from 100 square metres of bottom. At least two hundred different species were identified. "I know of no mixed-bottom beach along the entire Pacific Coast that can compare in sheer biological wealth to [this] one site," one astonished scientist wrote in his site inventory. For all its richness, the site is surprisingly fragile, vulnerable to unintentional damage from visiting recreational sailors or expedition kayakers. Simply to walk a shore that is silver with bleached shells is to destroy countless creatures with every step. Scuba divers and snorklers are of particular concern and a study in 1993 urged special protection and a strict management code limiting access.

Thirty kilometres north are two more zones of near-incredible biodiversity: Murchison Narrows and Faraday Passage, where a series of shallow, submerged reefs teems with different species in concentrations close to those found at Dolomite Narrows. That night, we anchor off Hot Springs Island where mineral water heated by geothermal activity deep in the earth's crust bubbles up into a series of pools, one of them on the cliff crest, where you can blissfully soak while watching the wind whip a blue sea into a froth of whitecaps.

The next day we push on to Windy Bay, gratefully accepting an offer from Shirley Wilson to sleep ashore in the new longhouse, perfumed with cedar and woodsmoke. When Chief Klue ruled here, the village was famous for the size of his house and a pole sheathed in abalone shells. The moss-covered remains of that house are visible today. Herself a granddaughter of chiefs—Skidegate on her mother's side, Skedans on her father's—this 53-year-old custodian is part of the Haida's Watchmen Program, begun in 1981. No, we're not intruding into her solitude, she assures us. She's already had fifty-nine visitors in the previous month. (Gwaii Haanas now records more than 20,000 visitor nights a season.) "I like meeting people. It gives me great pleasure telling people about our culture and our heritage, in taking part in preserving all this beauty for our kids and grandkids."

On her directions, we hike back into the bush to look at the lush, temperate rain forest that descends to the edge of the sea. Spruce trees five and six metres around at the base tower into a canopy far above. The ground below is muffled in thick green moss, the gloomy air motionless, yet alive with the insistent sound of water—seeping water, dripping water, trickling water. Nurse logs and snags are embroidered with the vivid yellow of witches' butter and sulphur polypore. Wet earth sprouts ghost pipes and coral fungus. And then, with spatters of rain riding a brisk wind out of the southeast and fuel so low the skipper says we'll be burning fumes when we cross the bar into Sandspit, we're forced to run for home.

Up past Kunga. Past silent Tanu with its overgrown house depressions and the grave of the great Haida artist Bill Reid, sleeping now under his new blanket of unadorned beach pebbles, waiting, perhaps, to be reborn into our world in some future child. On past Red Top Mountain, past Talunkwan, past the stark house posts and mortuary poles of Skedans, "Grizzly-Bear-Town" once so rich and powerful that the rancheries bore names like "Clouds Sound Against It (As They Pass Over)" and Chief Skedans' six-beam "House People Always Think Of." On Past Cumshewa. Past

Kundji. Past Sqe'na. Ahead of us looms the glow of neon signs and computer screens, modems and editors and policy wonks with cell phones debating the extent of protection this seascape of marvels deserves.

Behind us, grey layers of cloud and a grey sea, the once-sharp line of the horizon lost in murk and uncertainty. Somewhere back there, Xa'gi sinks into night. And with the gathering darkness behind and the habitation glow ahead comes an ineluctable sense that Foam Woman's cosmic doorway between the world of dreams and the world of reason is somehow sliding shut once again.

Fish Story

One by one, we cast off our moorings in the past that made us. Nobody knows the immediacy and power of this process more intimately than Norman Safarik.

Solid and stubby as a marlin spike, ruddy-faced, he rocks back into a squealing chair and swivels away from the desk. His head cocks like a man listening to something in the distance. I think I know what he hears: the tide of his life's work withdrawing down the shingle, leaving him beached on a strange shore after three generations in the same business in the same place. Norman's life is an eight-decade chronicle of Vancouver's unofficial history.

I get a sense of its breadth and depth myself one morning in 1992. I've gone down to the grubby eastside docks at first light, looking for those first impressions that are best served by the cold advance of dawn. The Campbell Avenue fish wharf is not yet demolished. To find it you sweep past the grim warehouses where my own father-in-law covered the Pier One riot more than half a century before, past the lifeless regularity of the container port, across the Heatley overpass and into a jumble of weathered sheds in the shadow of sugar refinery towers. Norman's father fled the decaying Austro-Hungarian Empire for this, escaping conscription into Franz Josef's corrupt, incompetent army. He started with a handcart at the foot of Granville Street, selling fish to Gastown out of a barrow before moving farther east. He found his moorings where I stand, watching the light spill westwards, peeling darkness from the North Shore mountains. There's a whiff of snow in the air. The moving boundary of light runs down the black bulk of the lumber carrier *Haida Monarch*, bigger than his whole establishment, and along the rain-slicked pier. Broken windows grace the abandoned Marine View coffee shop, another navigation buoy cut adrift. Beneath it, a sign of the times: "Ice plant closed. Will not re-open!" Along the pilings, where the double-ender fishboats would tie up and outside them the skiffs of the old Scots who would row out to Spanish Banks to catch plump morning herring for the kipper market, the bumper logs are eighty feet clear and the span of a big man's arm through at the base. A seedy air of decay belies the richness of history. Gulls pick at the huge, eyeless head of a ling cod. A discarded five-foot sturgeon rots on the mossy timbers below. Everywhere is the debris of abrupt departure—rope ends, empty tubs, paper, splintered boards, the layered smells of offal and slack tide.

The son built the father's hand barrow trade into a successful company—small, but feisty enough to survive in a landscape dominated by leviathans that grew fat cannibalizing their lesser brethren. As I watch, a grandson named Howard wheels and deals on the telephone with the high-voltage delivery that characterizes a business where the commodity is perishable, the supply insecure and the prices as volatile as the weather. Norman seems tired and distracted, his legendary toughness eroded by countless letters to bloodless bureaucrats, the rough edges of a life on the docks worn smooth by phone calls to politicians who respond with bagmen instead of solutions. I guess that the last thing he wants to do is talk to yet another wet-behind-the-ears writer about the imminent arrival of the future—a future that separates him from the waterfront and consigns a life's work to the dead ledger of history. The desk is of a kind you never see anymore, itself a piece of history. Hewn from planks of solid oak, it's scarred by cigarettes, battered by generations of hard, productive work. Buried under a midden of invoices, orders, letters of intent and contracts for supply, it squats in the back of a dark, scuffed cubbyhole above the silent sheds of Vancouver Shell Fish & Fish Co. Ltd. Dreary federal government posters display the varieties of fish and molluscs taken commercially on both coasts.

My guess is right. Norman Safarik is weary beyond any sense of betrayal, beyond anger, beyond the sudden, silent tears that welled up out of the deeps at a family dinner. They amazed his son Allan, the only member of the clan not directly involved with British Columbia's ailing fishery. Allan is my intermediary in this meeting. He's driven the hour in from White Rock to help me find a way through the bitter rhetoric that follows any prolonged argument with bureaucrats who owe allegiance not to communities but to invisible masters at the other end of the country. His father's leonine head swings back, fixes me in a level gaze. After fifty-three years on the dock, the old man's lease has been terminated, the wharf condemned. He's being kicked out of the landscape of his own memories. He's found new quarters for the company over on Terminal Avenue, but what kind of fish company operates out of sight of the boats and the fragrance of the saltchuck? And what kind of seaport, born with the salmon fleet, doesn't have a working fish wharf at the heart of its urban waterfront?

"Ask him about the time he sold BC Packers their own fish." Fuelled with paper cups of that truly awful, truly perfect coffee you only find on industrial work sites, Allan prompts from the wings. Poet, editor, publisher for a while under the imprimature of Blackfish, he has an instinctive sense for the story. Proof of that comes from the background. There's a pause in the swirl of employees coming and going on the business of Vancouver Shell Fish & Fish Co. Ltd. Everybody's waiting for Norman's answer, for the story. And that perks him right up.

"Ah, what the hell, he's dead. They're all dead. I guess you can't libel a dead guy," he growls. "It was the summer of—I can't remember the hell which summer—but there was going to be a real shortage of silver-brights. I knew it. The fishermen knew it." The only people who didn't know it, he says, were up in the management suite at BC Packers. They were distracted by corporate politics, embroiled in a joint takeover of Todd & Sons, another fish company formerly owned by British interests.

"The brass at BC Packers were playing games with the market," the old man rumbles. "They were worried that the hotshots in management at Todd & Sons might come in and get their jobs. So BC Packers wouldn't buy fish from them. Todd had millions of pounds of fish in cold storage that they couldn't sell." Farther down in the organization, there was much bitching about the quality of fish going out of BC Packers, he says, and much admiration for the silver-brights going out of Vancouver Fish for carriage-trade restaurants uptown.

"I looked at their fish, dark fish. I'd shake my head. You should be canning that fish. You don't want a reputation for selling that stuff. I can sell you some real nice silver-brights..."

They came over to look at his fish. "Those are the nicest fish you've got there. We're paying 40 cents a pound, your fish are worth 42 cents—get us 5,000 pounds so we can get a real look." He didn't tell them he was buying the silver-brights at cut rates out of the two million pounds in cold storage at Todd & Sons, which BC Packers now owned, and selling it back at a premium rate. When they found out, they sent an emissary.

"I was unloading this halibut boat and I could see there was a problem just by the way he walked down the dock. He said: 'Safarik, you're a blankety-blank crook. I'm going to beat your head in.'

"I laughed. I said, 'You wanted fish, you wanted it bad. It wasn't up to me to tell you that you already owned it.' He's flailing away and I'm backing up. I didn't want to hit him, you know. I could've decked him easy, but I didn't.

"Then I saw Ken Fraser coming over smoking his cigar. He was the president of BC Packers and he said to this young fella, 'You better hope you don't land one of those or you're fired.' He put his arm around my shoulders and said, 'Son, I want to see you.'" The two of them went back over to BC Packers management suite.

"'Bring in the ice and the whiskey,' he said. 'You son-of-a-bitch! What are you doing to my boys?' Then he poured me a drink. Two or three drinks. He offered me a cigar.

"'I don't smoke cigars,' I said. He said, 'You do now.'

"'Safarik, you're going to come to work for BC Packers,' he said. 'I'll give you a $50,000 bonus.

"'I can't,' I said, 'My dad would go broke.'

"'I know,' he said. 'And I'll personally go over and nail the goddamned two-by-six across his front door.'

"'I can't," I said, 'My dad...'

"Old Fraser, he was a very tough guy. He looked at me for a long time. Then he sent me down to see the guy who wanted to beat my head in a few hours earlier. That young fella was kind of red in the face. He was so mad he could barely talk. He said he had an order just down from Fraser's office he'd been instructed to discuss with me. Then he passed it over.

"It was a purchase order. He wrote me out a purchase order for the rest of their fish."

THE SECRET OF MCBRIDE

The kid rode a beat-up CCM that was clearly a hand-me-down through a string of roughhouse brothers. This species of bike is virtually extinct but the few survivors are more than a match for trendy Japanese mountain models with gears and enough cable to open a telephone company. He rode over without hesitation when I hailed, as kids still do in towns small enough that street-proofing means looking right and left at the main intersection and everybody knows everybody else.

"What's the most interesting thing to see in McBride?" I asked.

His face screwed up in serious thought. I watched him tick off possibilities.

"You wanna go see Rainbow Falls." He pelted off after his mates.

I checked this advice with Bob Balcaen who runs the TV repair and appliance store at the corner of Fourth and Main. Television didn't even arrive in McBride until 1973 and the water system is "straight off the mountain—100 pounds head pressure. The great Gods from the Ministry of Health want us to chlorinate the water," Bob scoffs. "We've been drinking this water for forty-five years. Now they decide to save us."

Bob, you'll gather, doubles as McBride's biggest booster after mayor Steve Kolida: "Population 592 last census. Nope, make that 600—inside the muncipal boundaries. Six hundred, write that down." What did Bob think about the kid's advice? He mused on it. Cast a skilled eye on my slumping physical capabilities. "The falls, eh? Keep going on up. You'll know what you're looking for when you find it."

Off I went in search of McBride's most interesting feature. Not the new arena, not the new fire hall, or the new high school or the renovated hospital—all the solid civic accomplishments that town fathers usually direct you toward. Directions from the kid on the bike led me through a stand of alder—old-timers will tell you alder likes its feet wet—and pretty glades enriched by the dying bloom of wildflowers and the white gleam of mushrooms.

The falls were a disappointment—huge in the imagination of a ten-year-old, barely a splash for someone fresh from Athabasca Falls. On the other hand, who could ask for a more restful place to pretend 8,000 kilometres of road wasn't waiting to be driven. A frothy lace of white foam spilled over a lip of limestone and twisted into endless patterns on the bedrock—a pretty display of subtlety rather than the awesome power of nature. I stretched out on the bank and counted birds darting through the foliage. Light airs feathered the leaves around them and white clouds tumbled overhead.

Then I remembered Bob's cryptic instruction and scrambled around the little falls and headed upstream.

The creek bed was too rough, so I worked up the ridgeback and tracked along the mountainside before descending a convenient chimney to the canyon bottom. Green shadows filled a hidden grotto, the kind you dreamed of after stealing late-night chapters from childhood books like *King Solomon's Mines*. Far above, wind roared across the cleft, hissing through treetops. The stone chamber rang with an eerie music, a curious, resonant piping sound like random notes teased from stone pipes.

Around the corner, sweating over deadfalls and tangles of roots, I found the musician.

It was a single stream of water plunging the height of a fifty-storey building into a clear pool. The source of the music was a series of deep hollows in the limestone ledges at the head of the falls. They trilled as the high-pressure jet passed over them.

The wind had swept the sky clean as a whistle and suddenly, a single shaft of sunlight lanced into the canyon. Then I saw what Bob Balcaen and the kid on the CCM bike wanted me to see. The whole hidden valley shimmered like a prism. Light glittered on vapours drifting on the grotto's still air. Over my shoulder the white column of water blossomed like a bride's veil made vivid with garlands of flowers. From the foot of Rainbow Falls, to the organ music of the earth, I stared in astonishment at McBride's most prized civic monument and considered the kind of marvellous community spirit that would direct me here instead of to the new hockey rink.

So Mad He Shot the Bar

Lost in the sun-drenched empire of grass where the Nicola Plateau straddles highway 5A from Kamloops to the canyons of the Tulameen, you'll find one of my favourite hotels. The Quilchena's ads pitch to the BMW set that comes with Ping golf clubs stowed in the trunk, dressed-down sow's ear affectations and genteel silk purse expectations. So the brochures shrewdly promise both "unpretentious hospitality" and "elegance." In other words, the cattle company cookhouse-come-coffee shop is balanced by the Swiss chef in the dining room.

Never mind all this marketing chaff. The Quilchena Hotel is also steeped in an earthy, big-knuckled past that makes it one of the most interesting windows on the province's rich and complicated story.

Its porticoed, neo-regency façade includes a veranda and white-pillared balconies. Varnished board floors creak satisfyingly when you walk on them. Ornate period furnishings and real bullet holes adorn the saloon bar. The building itself breathes history. This is a genuine working survivor from the days when British Columbia's cattle barons ruled the high country like feudal nobility and remittance men who'd found themselves as working cowboys from western ranches shocked Europe's horsey set by winning the world polo championships. In fact, the golf course where the hotel's summer guests play was once the polo field where teams from Kamloops, Walhachin, Ashcroft and Westwold competed before World War I. They made the best cavalry in the world and one of them, Gordon Flowerdew of Walhachin, won the Victoria Cross for a fatal charge into the machine guns at Bois de Moreuil in 1918. Ironically, that was the same year the Quilchena Hotel closed, killed off by prohibition and the replacement of the horse by the automobile. It wouldn't open again for forty years.

The Quilchena was built in 1908 by Joe Guichon, who kept it as his residence after it closed to paying guests. He and three brothers had headed north from California with the stampede up the Fraser to the Cariboo in 1857. The gold the Guichon boys found wasn't placer nuggets: it was bunch grass higher than a pack horse's belly. Driving south from Kamloops on the old road to Merritt, grateful for the sweeping efficiency of the new Coquihalla rocket-sled run that seduces all but local traffic, you can still see the sweep of the country that staked its own claim in those stampeders' hearts so long ago. You begin in the old community of Knutsford, looping past the Running Horse Ranch and Rose Hill Road, then skirt a chain of long, narrow lakes. Roll down your window, drive slowly and stop frequently to enjoy the vast panoramas and the ducks taking flight, leaving silver rings on glassy reflections. I interrupted a pair of heavy-duty bikers who had strayed from the highway and stopped at a gravel pullout to take in the view. One had shed his Harley-Davidson leathers, the modern-day chaps draped over the sissy seat. He spread his arms and his bare chest to the wind and did a small, pale pirouette of pleasure. Behind him, redwing blackbirds whistled from the bulrushes and flying clouds dappled the rangeland with patterns of sun and shadow. Everywhere here the skeletal underpinnings of the earth show through the thin mantle of topsoil, ridges and outcrops that look like nothing so much as the fossilized vertebrae of huge, extinct creatures, a reminder of the precarious grasp of our present life. Ribs of buckled rock stand exposed to the wind. Yet the harsh contours of the land are softened by the blowing mantle of grass that greeted those failed prospectors. In that wild grass they saw the future.

The first inkling had come less than two months after the first stampeders to BC's Interior had crossed the border and headed up the Fraser. General Joel Palmer crossed at Osoyoos, driving a herd from Fort Okanogan in the Inland Empire of what was not yet Washington State to supply the mining camps with meat. Talk about cowboys with bandanas and big hats driving vast herds of cattle and the first Hollywood image that springs to mind is John Wayne and the Santa Fe trail, but in fact that romance is as much a part of BC's varied history as it is anywhere else in the Old West. From 1859 to 1870, about 22,000 head of cattle were driven to the Barkerville mines, some from the mouth of the Columbia River—an overland trek of 1,800 kilometres through wild country with no roads. When they got to the Cariboo, fresh beef commanded the then-outrageous price of 50 cents a pound, which meant that selling steers for slaughter was a much faster way to get rich than sluicing colour out of gravel hauled up from the bottom of some cold, dank drift. Three American storekeepers, Benjamin, Abraham and Isaac Van Volkenburgh, hooked up with butcher Edward Toomey and a pair of experienced cattle drovers from California, Jerome and Thaddeus Harper, to supply meat to the camps at Barkerville, Richfield and Cameronton and soon were involved in one of the wealthiest enterprises in BC's Interior.

A typical herd of four hundred steers, fifty milk cows and fifty horses would travel up the old Fur Brigade route on the west side of Okanagan Lake, then veer west along the South Thompson to Fort Kamloops, Ashcroft and Cache Creek before heading north along the Bonaparte River until it could hook up with the old Brigade trail to Fort Alexandria and Quesnel. One of these Fur Brigade trails ran from what's now the Princeton area through the Tulameen and Nicola valleys to Kamloops. What we call the Nicola today was originally the country of Hwistesmexe'quen, Walking Grizzly, a famous head chief of the Thompson Indians between 1785 and 1865. Fur traders who had trouble getting their European tongues around the aboriginal syllables called him Nicholas, which Thompson Indians, who had similar problems with the European syllables, promptly changed to Nkwala. From this convergence of linguistic inabilities the place name Nicola emerges.

It was on the extension of this trail through the Nicola Valley, linking Hope to Princeton and the Similkameen Valley beyond, that the fifteen-year-old daughter of an early settler encountered a Hudson's Bay Company supply train in the early summer of 1860. Among all the historical accounts in the archives, Susan Allison's indelible recollection of a soon-to-vanish phenomenon endures as one of the most vivid and charming. She had gone for a stroll on the trail and stopped for a "feed" on the huckleberries that were ripening.

> I heard bells tinkling and looking up saw a light cloud of dust from which emerged a solitary horseman, the most picturesque figure I had ever seen. He rode a superb chestnut horse, satiny and well-groomed, untired and full of life in spite of dust, heat and long journey.
>
> He himself wore a beautifully embroidered buckskin shirt with tags and fringes, buckskin pants, embroidered leggings and soft cowboy hat. He was as surprised to see me as I was to see him, for he abruptly reined in his horse and stared down at me while I, equally astonished, stared at him. Then, as the Bell Boy and other horses rode up, he lifted his hat and passed on. I never met him again but was told he was

a Hudson's Bay Company officer in charge of the Coalville train and that he was never more surprised in his life than to see a white girl on the trail—he had lived so long without seeing anyone except Indians.

As the fur trade gave way to gold rushes that went up the Fraser, through the Boundary country and into the Kootenays and up the rivers into the Cariboo, the Cassiar and eventually the Klondike and Alaska, the Brigade trains gave way to cattle drives.

Following the Palmer drive came the Jeffrey brothers of Alabama, Major Thorpe of Yakima, Lewis Campbell and John Wilson, then Ben Snipes, William Gates and William Murphy. Daniel Drumheller drove a big herd west from what's now Alberta and so did Aschal Sumner Bates, not to sell but to ranch. Small wonder that many of these cowboys, seeing the open grasslands with high quality forage and plentiful water and ready markets, were quick to abandon the bone-wearying work of the trail rider for their own ranches. Today about 500,000 head of beef cattle can be found grazing on the 10 million hectares of ranch land that began with a few homesteads. The Nicola was a magnet for the first ranchers. Its lovely, rolling prairies had long been a favoured spot for overwintering livestock for the herds and pack trains bound north to the gold camps. The first pre-emptions were filed in 1868 and the first major cattle herd came in with the Moore brothers the same year.

Joe Guichon had first passed through the country in 1864 as a sixteen-year-old wrangler on the drives and pack trains supplying Barkerville and soon came back to homestead. Before he was finished he was running the biggest stock ranch in the Nicola Valley, a third larger again than the legendary Douglas Lake operation to the east. He built the Quilchena Hotel to replace a rickety bucket of blood he'd bought from Edward O'Rourke in 1904. If his hotel had some pretensions, it was still the heart of a working cattle ranch. Which leads us to those bullet holes in the bar.

The story, as told for the BMW set, goes something like this:

Toeless Smokey Chisholm, a ne'er-do-well living over in Aspen Grove, took umbrage at Constable William Fernie's arrest of the Grey Fox, as train robber Bill Miner was later filmified, romanticized and otherwise glorified to sell tickets at the box office. Miner had been living the quiet life after his release from San Quentin in 1904 after a career in the United States robbing stagecoaches. He got the itch again and with two companions bungled the holdup of a CPR train at Ducks, which we now know as Monte Creek near Kamloops. Constable Fernie nabbed him in short order near Chapperon Lake, which is about halfway from Quilchena to Kamloops. One cold snowy night, the story goes, Smokey rode over the mountain, walked his horse up the steps into the saloon and, to show displeasure with the upstanding citizenry, emptied his revolver, several bullets from which remain in the bar. There is a minor chronological problem here, since the Quilchena Hotel which now occupies the site was built in 1908, several years after Bill Miner was caught and sentenced to twenty-five years in the New Westminster Penitentiary, ostensibly upsetting Smokey. Miner escaped from prison in 1907 and fled to the US where he was promptly arrested again and sent to prison. He died in Georgia in 1913.

Much more likely, and in keeping with the BC Interior's gritty cowboy roots, is the story Richard Tenisch told me. Richard, late of Switzerland and fresh from running a restaurant in Banff, told me he got the real story of the bullet holes from Jack Patterson whose dad was punching cows at Quilchena when it happened in 1912. Old Man Patterson saw the shooting. Here's how it went. A Douglas Lake cowboy named MacGillivray was lubricating himself as he drove a small herd to the stockyards. He was well and truly drunk by the time he got to the hotel. MacGillivray got one drink from the bartender because he was so surly and then he was refused a second for the sake of his own safety. That made the cowboy so mad that he shot the bar. The only casualty was a barrel of rum. What happened to MacGillivray? The story doesn't go that far, but if he was anything like the other cowhands who woke up hungover in the root cellar after spending their wages in the saloon, he came begging for credit the next morning to pay for the hair of the dog that bit him.

Joe Guichon gave no credit. He was famous for not giving credit. Hungover cowboys were permitted to earn their morning drink by hand-pumping the water into the tanks which supplied the hotel guests. To pay for a barrel of Joe's Hudson Bay rum, MacGillivray must have pumped for a long time.

Spirit of the Stream

I find Wendy Kotilla at the end of four hours of greasy switchbacks, all negotiated with one eye cocked for loaded logging trucks and the other for unexpected washouts. The gravel leads to a pair of trailers in the salmonberry canes. Ponchos and waders hang above the dank duckboard between. A hard outer coast rain drums relentlessly on the fibreglass sheeting above. One trailer houses the cookshack. The other is a row of monastic cells, bare mattresses, a candle set out beside each for visitors. A small puddle of yellow light spills under the door of the last cell—Wendy's place.

This is the science base at Carnation Creek, a small watershed on the remote west coast, midway along Vancouver Island. It's typical of the small salmon streams that drain the British Columbia rain forest. Two forks of the creek and nine unnamed tributaries extend about eight kilometres back into the rugged, rain-swept mountains, channelling the runoff from 10 square kilometres into a main stem that once supported strong returns of chum, coho, steelhead and sea-run cutthroat trout. What makes this system unique is that it has been the subject of research into the impact of forest practices on a riparian ecosystem for almost thirty years.

Different sections were logged in different ways and the consequences were evaluated in the longest running fish-forestry research project in North America. It has proved timely and important research.

Across British Columbia a fierce debate rages over whether clearcut logging is responsible for flooding, erosion, land movements, silted-up fish spawning beds and muddy Vancouver drinking water, or if these events were going to occur anyway, forest cover or no forest cover. On one side is the "slides happen" school, arguing that most land movements have little or nothing to do with logging. Proponents of this position like to cite a three-year study in the Seymour watershed, which supplies Greater Vancouver's drinking water. It concluded that logging was a contributing factor in just five of 1,200 landslides. Such patterns are normal in the natural cycle, the argument goes. Landslides can actually be good for the environment, contributing to biodiversity by creating openings in the coniferous canopy for deciduous trees that wouldn't otherwise get a chance to flourish. The counter-argument points an accusing finger to the west coast of Vancouver Island where countless creeks like Carnation rise and tumble to the sea. Out there, says the Sierra Club of Western Canada, even usually cautious provincial environmental authorities are now describing recent slide damage in logged areas across the region as "incredible."

One typical example can be found on the upper forks of Carnation Creek. I climbed down into its deep canyons from a logging spur. It had been hacked across a clearcut hillside with 70 to 80 degrees of incline. Frankly, it doesn't take rocket science to predict what will happen on a slope like that when it gets several hundred centimetres of rain per year. In less than a kilometre of creek bed I saw landslides where hundreds of metres of debris had poured into the creek. I saw banks unravelling and a stream bed so choked with new blowdowns it was virtually impassable. This is all relatively recent, the result of logging that took place in the mid-1990s, long after the bad old days were supposed to be over. True, the rainstorms that triggered this appear to have been exceptional. But was it really a once-in-a-hundred-years event that won't happen again before the end of the next century? Possibly. But what are the implications for British Columbia's clearcut watersheds if it turns out instead to be the harbinger of an emerging norm as global warming changes precipitation patterns?

Among the important findings at Carnation Creek was the degree to which logging on steep unstable slopes above tributaries made the watershed vulnerable to damage from erosion. Landslides and debris torrents filled pools and destroyed fish habitat. Tree trunks and stumps scoured the stream bed, created log jams and forced the creek to change direction, further eroding banks and increasing sedimentation. The most dramatic damage followed a 1984 storm that dumped about 260 millimetres of rain into the watershed in twenty-four hours. What scientists call "maximum instantaneous stream flows" roared through at volumes of 65 cubic metres per second. But the record shows that rains almost as heavy as this now occur frequently. By 1994, about 25,000 cubic metres of debris had entered the creek since logging began. It was to study just such consequences that the remarkable long-term study of the impact of logging on a typical coastal watershed was begun. So far it has generated 154 scientific papers. The Carnation Creek project has since been scaled back dramatically but baseline monitoring and research continues. The study has vital lessons for our accelerating urban and industrial modification of landscapes from the Lower Mainland to the north coast.

For six years Wendy served as the guardian spirit of this remote outpost of science and solitude, although she refers to herself in more sanguine terms as "the hippie chick who went fishing and decided to save the salmon." She lived in isolation for much of the year, recording weather and stream levels, maintaining the traps where fish are counted, collecting biopsies from dead spawners, repairing the trails and bridges. It's cold, arduous work that demands meticulous attention to detail as well as initiative in everything from carpentry to survival. When a razor-sharp machete sliced into her arm, severing the tendons, she got herself out of the bush and to a MedEvac.

That's a long way down the road toward self-reliance from the girl who was pregnant at seventeen and soon a single mom on welfare. "I'm proud of what I've done with my life," she says. "I'm proud of what I'm doing for the world. I am a real role model for those young women." She's referring to the school groups which began turning up in increasing numbers for an educational tour of Carnation Creek. It's something Wendy worked up herself. In 1993 she shepherded three classes of kids from kindergarten to grade seven. By 1995, it was up to nineteen. She's since gone on to other things, pursuing formal studies in ecology and land reclamation at university, presenting her scientific papers, working on a kind of cultural biography of the watershed that reaches back to the pre-contact knowledge of aboriginal people who call it home.

The day I walked the stream with her the sleet was hard and the water rising. A flash flood had ripped up her trail boardwalks and we struggled against the current, clambering over slick log-jams while she told me about the debris torrents that originated in three upstream clearcuts. She pointed to a log, "Moved a metre in the last big rain." And a sandbar, "Its nose is scouring out and going into that pool." Like a cancer invading a healthy body, pea gravel now marches down the little river. It fills the trout pools and wrecks the salmon spawning grounds. Thick layers of silt signal eroding banks. The channel wanders erratically. Before logging, 4,000 chum spawned here. This year she counted just over one hundred. The coho were down to six males and one female in 1994. The cutthroat trout are showing deformities. The ancient race of trout trapped in the upper river since the last ice age is now dwindling.

She admits that what began as an exercise in reason and

scientific method has evolved into something spiritual, emotional, powerful enough to evoke poetry and meditation. But science always came first. Every day she shrugged into chest waders and hiked that lonely five kilometres up the stream bed, mapping the movements of log-jams, noting the painted stones with metal pins that chart the movement of gravel. Did I say lonely? Wendy always had the creek for company. It still talks to her all the time, she says, tells her things about itself—and about herself, too. For all the scientific papers, this woman has achieved perhaps the most intimate relationship possible with the complex ecology of Carnation Creek. She knows every mood and nuance. It's a knowledge that seeped into her with the rain and spray and long nights of solitude in the lantern light while the moving water sang outside her window.

Nulli Secundus

It is difficult for me to think of Dr. Margaret Anchoretta Ormsby without the image of her book-lined, light-filled study crowding into my memory. After all, it was there that she shaped so much of the thinking from which her public persona would eventually be forged. Her career as a scholar led to France, to the United States, to Canada—as she was fond of saying of all that lay east of the Rocky Mountains—and, of course, to The Big Smoke, as Vancouver is still called in the parts of this province that lie beyond suburbia. But the house in Coldstream with the exotic snowball tree on the walk and the Jersey cow in the pasture, the backdrop of crumpled hills with horses whickering in the cottonwoods, the white postbox with her father's name still unchanged twenty-five years after his death—this was her anchor, she said, the earthy centre that would always define her place in the world. As gracious as her literary style, the grande dame of BC scholarship sounded surprised on the telephone when I called to make my appointment, but welcomed an ink-stained purveyor of what she called "history on the run"—for me an homage; for her a collegial opportunity to catch up on the political gossip.

Margaret came out to greet me at the end of a gravelled drive under a leafy canopy. Flowers from another age splashed a riot of colour across the garden. Behind the trim house, with its cool veranda fronting Kalamalka Lake were five rows of working fruit trees; not the cloned dwarfs of industrial agriculture, but heritage trees. In my recollection, her wintry blue eyes take quick, shrewd measure of the visitor before her face crinkles into a vast, sunny smile. She is eighty-three.

History by BC's foremost loyalist has always explored how we are stitched into the fabric of the larger Confederation. Bloodlines, family connections, the mysteries of the heart as much as of commerce, create an unseen web that links east and west in ways the demagogues of division and regional parochialism seem not to recognize. "This house was built in 1909 by a United Empire Loyalist. He was married to one of the Buells who founded Brockville. Here we are in BC but this is a typical St. Lawrence River house." Indeed it is, even to the boathouse with sleeping lofts. In the kitchen, copper-bottomed pots frame her wood stove—a two-burner Gurney—and the collection of ceramics. They are stunning pieces by Alex Ebring, western Canada's first potter and a friend of Ormsby's mother, Margaret Turner Ormsby.

And up a narrow, twisting staircase that led to the small dormered room upstairs overlooking the tranquil green waters of the lake, is her study. A solitary table stood in the centre of a lustrous hardwood floor. Around it were scattered the books demanded by a questing and eclectic mind: *The History of the Reign of Ferdinand and Isabella; Woman as a Force in History; Pioneer Reminiscences of Puget Sound; The Masters and The Slaves: A Study in the Development of Brazilian Civilization.* The study itself is a remarkable piece of British Columbia history. It was added to the creaking, comfortable house designed for Quebec's townships in the year she was born in a log hut. The old scholar sat at her table in the warm light and reminisced.

She told of first visiting the house as a little girl in the light-hearted Okanagan that existed before 1914. She'd fall asleep while her father and mother, George Lewis and Margaret Turner Ormsby, danced all night before travelling back to Lumby and the general store they kept there. In the telling, her eyes would sparkle like a five-year-old's. Then they'd darken. "It was very gay before 1914. Then came the war. We spent the winter of 1914 in Victoria. Dad was in army training at The Willows. In fact, we travelled to Victoria on the boat with the troops and I remember the submarine escort. I always had great feeling for Premier McBride and the submarines."

On the eve of war, Victoria and Vancouver bristled with rumours of espionage in the large German community, and of cruisers from Admiral von Spee's China squadron poised to attack the naval base at Esquimalt and sink shipping in Vancouver harbour. The city's foreign-language press hadn't helped things with sensational stories about crushing German naval victories. During the last days of peace there was absolute certainty in government circles that an attack was imminent and that citizens of German and Austrian descent would support it. When Richard McBride learned that a Seattle shipyard was building three submarines for the Chilean navy, he knew that when war was declared American neutrality would lock them up. So he simply wrote a cheque on the British Columbia treasury for $1.1 million and bought them in anticipation of the navy's desire to do so. The submarines slipped out of Puget Sound under cover of darkness and waited, five miles outside Canadian waters off Trial Island until war was declared and then proceeded to Esquimalt. Their appearance off Victoria on August 5, 1914, caused a panic and they were in danger of being shelled and sunk as German invaders by the shore batteries at Fort Rodd. But cooler heads prevailed, the submarines were turned over to Canada and BC's navy ceased to exist after three short days.

The war, unfortunately, would last much longer. Of the 43,000 men from BC who went to fight for the Empire, almost 20,000 would be killed or wounded. "I remember my father coming home in 1917," she told me. "My mother took me down to the station at Vernon to meet him. I had this awful feeling that I wouldn't know my own father. But I knew him all right. My father was in the first gas attack at Ypres. He was wounded eight times on the Somme. I only really knew him as a disabled man. But he never made much of his wounds. He was terrified they'd keep him in hospital." George returned from the Great War to work as a scaler in the bush while forty acres of fruit trees matured. He educated three children on a labourer's wages. And then he bought the Kalamalka Lake house in 1946. Fittingly, his daughter died there, peacefully—filled with tranquility says her nephew James Marcellus—on a grey November Saturday in 1996.

When time closed Margaret Ormsby's full and remarkable life at the age of eighty-seven, it severed for the rest of us a vital connection between urban British Columbia and its rough-hewn roots. It is a connection that the UBC professor emeritus of history increasingly believed a self-absorbed and introspective Lower Mainland risked losing at its peril. For Vancouver and Victoria to abandon their sense of community within the vast historical geography of British Columbia, she said, would be for the province to lose its soul. And she believed the same for her province's commitment to Confederation. She had little time for what she saw as separatist flirting among political opportunists. "I want Canada preserved," she said the last time I talked with her before her death. "BC is Canadian. I do not approve of any spurious northwest union. There are ties that extend across the country from east to west." She, after all, had spent her life's passion attacking intellectual parochialism while arguing the importance of linking the study of regional history to its global context.

Her parents travelled by covered wagon to homestead in BC's

rugged and remote Interior. Their route was the stagecoach road built to enable the rush to spoils in the golden Cariboo. Margaret, born along the way on June 7, 1909, was thought to be the first non-aboriginal child born in the raw, often brutal mining camp of Quesnel. For a time she lived in a tent. For a time in a rude cabin. In winter, her parents broke the ice on the frozen streams to bathe. Margaret took delight in regaling visitors with the fact that she had become a distinguished graduate student before electric lights came to her home. Any reading in her father's library was done by the buttery glow of coal-oil lamps.

She grew up in an unusual household for her time. Her mother thought every woman had an obligation to be economically independent. And, in an age of corsets and coy manners and easy contempt for suffragettes, George Ormsby believed that women were not born to be either ornaments to male pride or household drudges. He believed that God gave women brains for a purpose. He illuminated his daughters' surroundings with books, ideas and vigorous discussion and he pressed his belief that the world of the mind should prevail. Coldstream had no high school at the time. Many settlers, products of British public schools now eking out a living on the frontier, couldn't afford to send their sons to their old schools. Boys went away to Vernon Preparatory. Girls weren't expected to find much use for higher learning. They might get a year in finishing school or learning etiquette with a French family.

"My father was very different. He thought that education was important and that women's education was exceptionally important. My mother thought that every woman should be able to earn her own living and I thoroughly agree." When Margaret Ormsby died, she held honorary degrees from six universities, a doctorate from Bryn Mawr and a distinguished string of academic achievements which included chairing the history departments of Sarah Hamilton School in San Francisco and at the University of British Columbia. More importantly, she was an intellectual force, almost single-handedly creating a national and a North American context for the province of BC within Confederation.

When she was appointed chair of UBC's history department in 1965 there were strenuous objections on the basis of her gender. Dr. Robin Fisher, acting dean of arts and sciences at the University of Northern BC was the last doctoral candidate she was to guide to a successful dissertation and he remembers it. "Even in the 1960s it was a really chauvinist institution," he recalls. "That was a tough role and she was tough enough for it. She was a very strong woman." She had to be.

"After I had my PhD I taught for six years with just a ten-month appointment, year to year," she recalls. "I did benefit from the fact that there was a great shortage of professors after the influx of men following the Second World War. That was when I got my break. I could be afforded. They could get me cheaper than anyone else. I was told this by the president of McMaster when he offered me a job. But I wasn't accepted as equal by some of my male colleagues, even when I was appointed at UBC in 1965. 'You don't think like a man,' they said!

"I'm not an extremist. There are many people who would not call me a feminist. But I do believe in fair play. Women are not given a chance to show their stuff. You know, I think they are down to one woman in that department now."

Prior to 1960 just about the only place to obtain a PhD in history was at the University of Toronto, Fisher says. Under Margaret Ormsby, UBC elbowed its way onto the stage of national scholarship. Dr. Allen Tully, chair of the department at the time of her death, concurs: "Under her leadership the department quadrupled in size. She really brought the history department at UBC into the modern era."

John Bovey, another of her students and recently retired after a distinguished career of his own as chief archivist for the province, says it was her determination and vision that for the first time placed BC history in a national, North American and international context. "She felt that one of her major roles was the expansion of graduate studies. The number of theses she supervised is quite remarkable. And many people aren't aware of it, but she also taught medieval history at the same time."

The work which established her preeminence as a scholar, however, was *British Columbia: A History,* the first one-volume treatment of the province to be published since 1914. Bovey recalls that the book was commissioned by the BC Centennial Committee because it wanted some enduring legacy to remain after the celebration of the one-hundredth anniversary of the birth of the province from the union of Vancouver Island and the mainland gold colony in 1858. Margaret accepted the commission in 1956, then researched and wrote the history in fourteen months. The book remains an astounding accomplishment. It is an eloquent masterpiece of historical literature, astonishing in its scholarly breadth and scope and yet easily accessible to the lay reader. "It is not that easy to capture the history of the province in a single volume," says Fisher. "In some ways it is still the best, certainly it's the best written. She did have a great sense of the English language as an historian. That's something that's largely missing today. She put British Columbia's history on the map. From it developed an alternative to the Toronto view of Canadian history." "The book, in many ways, has not been surpassed," Tully agrees. "Her big work has stood alone for thirty years."

Although she retired from UBC to live in the Kalamalka Lake house in 1974, she maintained a vivid and abiding interest in presenting a non-Vancouver view of the history of BC. Even at the end of her life, the flame kindled by a wise and loving father so long ago burned as brightly as ever, a tireless curiosity still sifting through the unanswered questions of our past. She reflected on a lost generation, children of the European newcomers and aboriginal mothers: "They fell between cultures. What happened to them and where are they now?" She wondered about wealth: "I'd like to study the old money in BC—where it came from and where it went."

At the age of twenty-two, Margaret Ormsby had defiantly completed her master's thesis on the history of the Okanagan Valley. At the age of eighty-one, she returned to her scholarly roots with *Coldstream Nulli Secundus,* a meticulous, fascinating history of her beloved district, folded into the rolling North Okanagan hills. "In a sense, that closes the circle," says John Bovey. "It marks the end of an era in our history."

Out on the Swiss Burn

Buck Flats Road twists up into the Morice River watershed where Houston sprawls across Highway 16, halfway between Prince George and Prince Rupert. Exactly 22 kilometres down this road, what they call the Swiss burn cuts like a draftsman's line across the Nechako plateau. It strikes an unmistakable border between the land of the living and the empire of the dead. Late in May of 1983, with temperatures in the 30s, fire got away from two tourists smoking their catch on the Morice River—hence the "Swiss" burn. The flames flashed through pitch-laden stands of lodgepole pine and spruce faster than a man runs. In eight minutes it exploded from a campfire to 100 hectares. Two hours later it was 1,000 hectares and smoke eclipsed the sun over Houston.

On the day of my visit, rustling cottonwoods trace the river bottoms with sound, the noise of their leaves mingling with that of the current over pebbles and gravel bars. Here and there, the golden hayfields of dispersed homesteads reach back to the north through autumn forest. South of the fire line, a different world. A stark forest of blackened spars stretches off to the far limits of vision. More than half a decade afterward, standing on the running board of my truck, I can still chart the ferocity of the fire—great corkscrew swirls where hurricane winds inside the blaze flattened trees.

Frank Ebermann watched it come toward him, a wall of flames ten storeys high. Frank is an artist, a sculptor finding his painful way toward the wedding of function and form. He'd come up into the remote north country seeking tranquility.

"We saw all this heavy equipment coming down the road at a high rate of speed. Just fleeing. Looking down from the mountain, it was like Dante's Inferno. Smoke and the valleys boiling with flames. The burn made its own firestorm, uprooting the trees as it went." He grabbed his little girl Amai, found his wife Sofia and fled with the suppression crews. The fire took the grand log house he'd spent eight years building on the quarter section he'd put up everything to buy. It took all his work. It turned his beautiful retreat into a charred ruin: "All that winter, smoke from the hot spots burning in the ground. Dead animals everywhere."

Another person might have yielded to despair. Frank found liberation in his loss and faith in himself and the world around him. "This fire tore me away from any materialism. After the fire, my work began to change...Before I was a technician, then I became an artist. My whole philosophy changed. The fire enabled me." He sweeps his arm across the horizon: "We humans perceive it as ugly. It conflicts with our human desires. We want to own the beautiful scenery. But we never really own anything. This is not ugly, it's natural and nature is beautiful. We have to learn to see. That's what art is about—seeing. When the sun shines into the burn now, the spirit changes entirely. Outsiders come and they think it looks awful. Not to me." Frank carves big, swooping forms out of wood—the intensity of the forest fire dried the wood better than any kiln. Now he has a source at his doorstep, a gift of fate. He treks the burnt landscape searching for it.

Later, we take his canoe upstream, running it over the beaver dams and working up the riffles. We come to still pools beneath a canopy of willows, beach the canoe and strike inland across the Swiss burn. Frank is right. The sun slips between dead trees and the thin groundcover blooms into vivid, luminous greens. Underfoot, a carpet of blueberries, kinnikinnick, wild roses, raspberries, strawberries. Countless wildflowers perfume the air. I bend to a seedling lodgepole pine, cup my hands around the tiny tree, a green needle thrusting its way into the world through a black crust of ash. "Yes," says Frank. "The heat from the fire bursts the cones and sets free the seeds. They fall on this cleared ground. The lodgepole lays down new life in the midst of the fire."

New life springing up in the ruins—like Frank and his work, learning from the terrible beauty of what has been changed utterly.

Indeed, the whole story of how Frank and his family came to be here on this burnt-out plateau that's tagged Buck Flats on the local maps is one of love's power to endure.

Sofia tells it over a rack of moose ribs roasted in honour of her visitor from outside, a cold bottle of wine she can't really afford that meant a special trip all the way into Houston, all laid out on her best white tablecoth in a room that's luminous with her husband's woodwork. A short, sturdy woman with high, flaring cheekbones and flashing black eyes, I ask her right away if she is aboriginal, curious because she seems to be and not to be at the same time. "Yes," she replies. "Indian. But not from here. From the Andes." She is the daughter of an Indian woman who was abandoned by the Spanish gentleman who came to the New World to make his fortune—she snaps her fingers, dismissing him—and left everything of human value behind when he'd made his money and gone back to "civilization." It is a story that is not uncommon in these parts, either. So how did she come to be living on a wild quarter section of Canada with an artist? One of those things that happens between a man and a woman, she says. What did Tennessee Williams call it—the hidden power of love to burn you to the ground?

Frank had set out from Germany to cross South America on horseback, north to south, one of those Quixotic adventures upon which young men still embark. His horse died in the remote uplands of Argentina and he walked and walked and walked, all the way to the village of Malargua to try and buy another. The luckiest day of his life. Sofia's sister, the vivacious Violetta, invited the exotic stranger to talk to kids at the school where she taught. It seemed appropriate to invite him to dinner by way of both payment and hospitality. Shy Sofia was called upon to help with the cooking.

Like Desdemona, she says, she was swept away by this handsome young traveller dressed in black with his grand tales of adventure.

The next morning, a knock on the door. When she answered, it was the rumpled dinner guest. He said he hadn't slept. Sofia was horrified: "I thought it was my cooking."

"No," he said. "I couldn't sleep thinking how beautiful you are."

Unfortunately, in one of those arrangements sometimes made in families with many daughters, she was already betrothed—to the local police captain, "an old man, going bald under his hat." Sofia was no woman of the past. She followed her own heart. In the time of the disappearances, she decided they should elope, disappearing on their terms rather than those dictated by the captain's offended machismo. But it was too late. Sofia was arrested and jailed for the crime of knowing her own mind. She was released to serve as nanny for a general—only on the understanding that her romance with the European was finished.

Instead, she fled with Frank, leading him deeper into the high Andes. "We walked through the desert and filtered drinking water through his felt hat. We ate goats and the occasional small animal. We lived in caves in a valley full of condors and eagles." Arrested again, this time both were jailed.

"I saw things in prison that were so horrible. So horrible. It always makes me feel dirty thinking about them. So sad," recalls Sofia. In Frank's prison, "the rats were so thick I stayed up at night to beat them off with a blanket. But I was treated well, being German." Sofia was not. She was Indian.

"A woman is never a full citizen there," she says, leaning fiercely into her kitchen, the room so full of light yet crackling with her intensity. "To them an Indian is never a full human being. To them, an Indian woman is less than a child. They assume you are helpless and stupid."

The German embassy intervened to have Frank released but the deal required him to agree to leave the country. He agreed, then stalled his departure. Sofia was released and told he had gone home to Germany, to forget this waywardness. "But I knew he hadn't. I knew Frank wouldn't leave without me. So I went into the city and I found him."

And so she fled her country of condors and eagles for crowded Europe, where she and Frank were married two weeks later. "My sister asks if I won't go back to Argentina. I say I don't trust Argentina. I will never trust Argentina again. Anyway, I married my husband—I'm not married to Argentina."

They found their way instead to Buck Flats with an axe and not much else and made their home under the sweeping branches of a big northern spruce, sleeping in the swaying arms of the wilderness while they built their log cabin. Hunted out of Argentina for crimes of the heart, she says, they'd ceased to trust the values of civilization. They wanted now to build their own shelter with their own hands and live a life that was as self-sufficient as it could be. "This is so safe and friendly here," says Sofia. "Only the animals. You are safe with the animals—it is humans who can be dangerous." It took eight years to build the full log house he'd promised Sofia, finding the perfect trees, trimming the timber, moving them by human muscle and sweat. Yet when it was finished they shared a satisfaction that most people can only imagine. "I thought there was some purpose to my life. I had achieved my dream. I had my log house in the wilderness," Frank says. And then he furnished it with the work of his own hands—great sweeping contoured chairs carved out of single pieces of wood, stunningly beautiful tables that survive now only in photographs.

The same spring the house was finished came the fateful Monday of May 30, 1983, when the fire came raging up from the south. "You could feel the heat from the flames half a kilometre away and the noise they made was like the roar of a 747 jumbo jet." He remembers and his voice drops to a murmur. "It moved so fast, extremely fast. It burned all this in two days. Just amazing. I didn't know what a real forest fire could do." That night they prayed for rain, a switch in the wind, anything to keep the fire shifted away from the house. They had no insurance. They never imagined that an overnight whim of nature could take everything.

On May 31, they came back up Buck Flats Road and saw the same landscape of destruction that made me stop on my way in, climb on the running boards and stare. When they got to their homestead, "There was nothing. Just nothing. The house was gone. The bridge was gone. Even that tree that kept us in its arms when we first came. The power lines were gone. There was nothing but a smoking hole in the ground with a stone chimney sticking out of the charred rubble."

It was a story and picture for the front pages in Vancouver, 600 kilometres to the south—and then oblivion. One day's headlines for the suburban Lower Mainland; for them all beauty and security seared from their lives.

And then one day, out of the Swiss burn came a plume of dust and ash. It was a pickup truck. It led a ten-man work crew. They were big, square-handed men. Sun-burnt and cheery, but serious. The Mennonite Bretheren had seen the story published by the *Vancouver Sun* and come all the way from the prairies to lend a hand.

"Most of them were Alberta farmers. They came swinging their hammers and built us a new house. It went up real quick.

Three weeks later, we had a home again."

It's the same home I'm sitting in now, refusing a last glass of rhubarb wine because of the long drive back to Houston and Prince George. A home filled with light and love and creative imagination. Sofia rises and goes to the bookshelf, returns with a battered, dog-eared cookbook. Her eyes brim with tears.

"They gave us a quilt for the bed and this Christian cookbook. It was our housewarming present. I said: 'How can we ever thank you?'

"They said: 'We don't do what we do for thanks. We do it for Christ. Don't thank us. Do the same for your brother in need.'"

Pulling out for the drive back to Houston, with rain squalls and sun showers scudding across the uplands, I stopped for a last look at the awesome sweep of the Swiss burn, the terrible beauty of nature and its enduring power to transform. As I watched from the ridgetop, a single arc of rainbow seemed to drop right into the Ebermann homestead. The riches I had found at the foot of that shimmering arc are not mythical or a fairy tale. They are the greatest wealth of all: the gold you find in simple human kindness.

We Are Out There

Above the rotting Union Steamship wharf, finger pointed into the clag and spindrift hissing coastward over brushed-steel swells, a skipper in oilskins stands like an accusing prophet on a huge granite boulder. "We are out there," says the inscription at his feet. The bronze image is a laconic reminder of the human price exacted by Prince Rupert's bondage to the sea and it hints at the deep currents of grief and worry that course through its present. In this kingdom of grey, 750 kilometres northwest of Vancouver, the rain drums down 236 days a year and the deepest months of winter average only thirty-eight hours of sunshine. Lately, the town's prospects are as difficult to predict as a sunny day.

There is uncertainty about the future of more than 750 jobs at the town's big pulp mill, an economic engine whose influence reaches thousands of kilometres inland but which is on life-support from the provincial government. There's frustration over the ongoing dispute with the Alaskans next door over allocation of dwindling salmon stocks. Tourism took a body blow after the summer of 1997, when fishing boats blockaded an Alaskan ferry and the state retaliated by suspending service. Now it looks as though federal conservation policies and diminished runs will mean almost no fishing for the next two years. Prince Rupert's unemployment rate now hovers around 16 percent. At the Community Fisheries Development Centre where jobless fishers can sign on for counselling about skills upgrading, the average caseload per counsellor is 300 clients. Since 1978, 7,621 more people have left Prince Rupert than arrived. And yet, to paint a picture of abject despair would be inaccurate. The net population loss to out-migration has been more than offset by 9,392 local births, an affirmation of hope that won't be denied. Railway booms and resource busts, sea otters and fur seals, halibut and herring, the delusion of limitless timber resources and an endless bounty from the sea, grandiose dreams of commerce and industry—from its beginning, Prince Rupert's seen all of them ebb and flood like the ocean tides.

The town's founder, Charles Melville Hays, an American wheeler-dealer who made this the terminus of the Grand Trunk Pacific Railroad—his statue stands by City Hall—believed Prince Rupert would eclipse Vancouver. That dream went down with the *Titanic* in 1912 as he returned to the north coast from a business trip to raise venture capital in London. But Prince Rupert endured. The 151 ceramic plates set into an austere brick wall below the imposing statue are a testament to that. They fix in community memory the lost and the drowned and those who spent their homespun lives in the embrace of the cruel and unforgiving ocean that surges at Canada's westernmost gates. Here you pay tribute to the *Good Partner, Ocean Clipper, Queen of the North*. Here is *Combat*, disappeared in passage from Haida Gwaii. *Bravado* sunk with six hands. *Ocean Invader*, lost with the Parnell family. *Gustav*, capsized with John, David and Wendy Trail, all drowned crossing Douglas Channel.

"Yeah, *Gustav*. She iced up to windward. The force of the gale kept her upright. Then they come around into the lee of a headland and once she's out of the wind she rolls over. Weight of the ice. Just like that." Foster Husoy, seventy-five, fished the north coast for half a century and knows the stories behind most of these memorial tablets. In fact, he was a model for the statue above. On occasion, great Mother Ocean has taken his boats, his friends, his life savings. And yet, he says, she has also given him everything he has.

In 1965, with a crew of five and 42 tonnes of herring from the icy Portland Canal in his hold, Foster was running for home before heavy weather. Rounding Haystack Island, they turned into huge seas. The cargo shifted, the boat rolled over and sank, and Foster, who can't swim a stroke, found himself clinging to a capsized life raft. It was pitch dark and blowing 45 knots. Whitecaps were breaking over them. Then it began to snow. He mourned for his lost boat, named for his son Lance, "the most beautiful little boat you've ever seen in your life." He thought about the frightened men who were his responsibility. He thought about his wife Hilda, waiting at home. "Things weren't going very good for Foster. I've never been so goddamned cold in my life and I never want to be that cold again," he muses, running fingers over the memorial tablet for Harald Marelius Eriksen, *Viking I*, 1888–1965. "Then I thought, you can't afford to die, Foster. You're so far in debt that Hilda will never make it."

It's a thought that must be familiar to many on the strangely quiet commercial floats where 1,100 fishers are facing another disastrous season. Many are so broke they can't pay their licence fees. The runs are clearly in trouble. American greed, Ottawa's mis-management, bloody-minded sport anglers—there are plenty of targets for recrimination. But the bottom line is that you can't fish for what's not there and these days, what's there you're not allowed to fish for. Art Sterritt of the Northern Native Fishing Corporation doubts half the gillnetters in his outfit will fish next summer. "You're looking at an average of 250 to 300 sockeye per gillnetter and around 1,000 fish per seiner," says Joy Thorkelson, North Coast rep for the United Fishermen and Allied Workers Union.

"That won't even cover annual licence fees. This means fishermen for the next two years have no hope. This means fishermen are going to have their homes seized. They are going to lose their boats. And the people who rely on the fishing industry are going to be forced out, too." And if the boats don't fish, the shoreworkers don't process the catch. Last year people worked two-thirds fewer hours on the canning lines and at the filletting tables.

How important is fishing to this community's economic health? Over the last decade, direct payments to commercial fishers and wages paid to fish plant employees in Prince Rupert amounted to almost $1.4 billion. In 1996 alone, the total was $286 million. Because so much fishing revenue is recirculated in Prince Rupert—when the fleet comes in the grocery bill at the local Safeway can top $150,000—the multiplier effect in the local economy more than doubles the impact, estimates accountant Odd Eidsvik. But in 1997, economic benefits from fishing crashed to less than half their previous levels. And in the two years following, it might diminish by half that again, Eidsvik says.

Ron Tsuji, manager of Pacific Fish and Twine, sees the consequences every day. He outfits the average gillnetter, "all hung" with nets and floats, for about $4,000. "Look at the store," he says. "It's empty. That's how it's been all year. Normally at this time of year I have 100 to 150 nets ordered. This year I have three." Victor Prystay sees it too. He's worked as a mill hand, as a highliner on a halibut boat, and now he sells real estate from the Realty World agency he owns and operates on Third Avenue. "It's very tentative out there," he says. "Everybody is a little frightened about what might happen next. Our business has been hit extremely hard." The most dramatic evidence, he says, is found in the price range for starter homes which should be a hot market with interest rates so low—"But it's not there. It's less than 10 percent of sales." The cold statistics back him up. In 1997 there were only fourteen housing starts in Prince Rupert, small business incorporations plunged by

46.6 percent and business bankruptcies have almost tripled since 1994.

Des Nobels, 44, has been fishing from his 10-metre gillnetter the *Vonni Dee* for twenty years. He's blunt about his prospects. "I have no intention of ever going salmon fishing again," he says. "I've had my fill of the politics. I've always been a very conservative fisherman. I'm not over-capitalized. I own my boat. I own my home. I live on what I make, which is about $40,000 a year. Last year I made my expenses. It's tough. It's extremely tough. Nobody knows what they are going to do. You've invested your whole life in something. You planned to do it for the rest of your life. Then it gets cut short and there's not a hell of a lot out there to go to."

Foster Husoy must have felt just about the same way trying to keep his freezing deckhands awake on their overturned life raft thirty-three years ago. Then, six hours later, out of the blue, they were picked up by a passing tanker. The anecdote might be a parable for the complicated mix of grief and stoicism, despair and determination, hardship and hope that seem tattooed by the rain itself into this town's gritty soul. "Hey," says almost everybody after reciting the litany of woes, "this is a great place to live. Rupert has a lot going for it. We've been down before, we'll be up again." In that sense, Foster, long-retired from his 65-foot seiner *Bold Pursuit*, makes an apt metaphor for the durable, resilient, "never-give-an-inch" spirit of the north coast. The statue for which he posed is a powerful work of art in its own right, but it is also a surprising statement of sophisticated self-perception for a community as remote from the urbane Canadian mainstream as Yellowknife or Havre St. Pierre. For if the maritime memorial insists on remembrance, it also points to the course ahead, to a hopeful future beyond the heavy economic weather buffeting Prince Rupert. Many are convinced besides despair, prosperity lies "out there" on the eternal sea.

Behind the bronze mariner, ascending from the rockbound

coast, the town of 16,721 spills inland in a chaotic, medieval sprawl. The waterfront has a stark industrial beauty of its own, a tangle of CN humpyards, commercial moorings that go back to the fur seal fleet, sheet metal fish-packing sheds, modern ferry terminals and mouldering docks that juxtapose the age of steam with the age of turboprops at an adjacent seaplane base. Along the ridge behind, the commercial district straggles down Second and Third Avenues, an eccentric, exhilarating collection of styles and shapes that vary from the steel and glass of up-market hotels to the seedy, gold-rush rococo. Behind the ridge, past seething tidal rapids, Highway 16 snakes away through the cedar jungles on the wet side of the Coast Range toward a distant interior from which chip trucks rumble westward beneath pillars of steam that set the Skeena Cellulose pulp mill shimmering on the horizon like a rusty, tubular mirage. Back streets twist through rock cuts still scarred by the drill channels left by diamond bits. They crawl up the craggy jumble of metamorphic knobs and outcrops that signal the leading edge of Wrangellia, the crumple zone where continental plates collide with forces that render stone plastic and pliable.

These immense, tilted tectonics are what make Prince Rupert unique, provide it with the finest, deepest harbour on the western side of North America and then thrust that harbour 600 kilometres closer to Asia by sea than Vancouver. To put that distance in perspective, as the crow flies, Prince Rupert is as far west of Vancouver as Vancouver is from Jasper, Alberta. The economic implications are profound. On a turnaround to China, the difference is about sixty hours—that amounts to $60,000 in fuel savings, says Prince Rupert Port Corporation president Don Krusel—and that translates to savings of $3 a tonne in cargo costs for a 20,000-tonne grain carrier. And it only gets better. Lay a piece of string on a globe so that it connects the industrial heartland of America to the mainland of China and the line from Chicago passes right through Prince Rupert. This route offers the lowest gradient and the least curvature of any transcontinental railway. Less work against gravity and less friction on the rails means faster, cheaper passage. "It's slamdunk proven that we can ship it to the Orient way cheaper than you can ship it from Vancouver," says Victor Prystay. Which is why, just offshore, glittering by night like so many horizontal highrise towers, the great ocean-going freighters form a whole transient suburb of their own. More than 300 ships a year anchor in the harbour, loading millions of tonnes of export cargo. This affluent, cosmopolitan, eternally transient suburb brings its own downside of familiar urban problems—a steady market demand for casual sex and a corresponding vector for drugs and VD. But Prince Rupert's not the kind of place to let that slide, either. One internal report on prostitution prepared for the attorney general notes the mother who armed herself with the family firearm and "retrieved" her wayward teenage daughter from a ship.

For all the wrangling over government bailouts for the ailing pulp mill, for all the bitterness over federal fisheries policy and the impotent rage against Alaskan over-fishing, Prince Rupert remains a vital maritime city. Salt water pulses through its heart and the vast Pacific swings through its dreams of past and future. "The port is the most important single physical asset that the community has," says Mike Tarr, president of Northern Savings Credit Union. "Right now this is strictly an export port. If we can get imports, if we can get two-way traffic to the Rim—wow!" Don Krusel is working on just that. He sees a future in which ships carrying coal, grain and container traffic from the deepwater berths at Ridley Island will backhaul goods and products from Asia. In 1998, for example, the first major import shipment of high grade steel—70,000 tonnes of plate destined to be rolled into pipe for the northern Alberta oil play—moved through the port for the first time. That's 1,000 railway cars loaded with steel going to Edmonton. "The future is the port," Krusel says. "We are an undervalued asset. We are poised for dramatic growth." He sees potential everywhere—

pulp exports from the northern prairies, new coal mines coming on stream in Alberta, ethylene exports from new gas fields in BC's northeast, maybe a steel mill supplied by coal ships backhauling iron ore. "A lot of people are down because we've gone through a couple of tough years. But if you keep your eye on the long-range picture, you can't help but be optimistic," he says.

For many in Prince Rupert, however, the future is now and it is painful and filled with worry. After twenty-two years at Skeena Cellulose, Paul Lebedick got downsized along with twenty-three others. "My gross pay was $5,000 a month. Now I'm down to $1,300 and after August I'm down to zero. You know, I should have had a severance package of $130,000, but they made me another creditor. All I got was 10 cents on the dollar—$22,000. With $130,000 you can do something, maybe start your own business. You can't do much with $22,000. You can't lose faith in yourself. You can't lose hope. But it's hard. My house is for sale. Soon there'll be no money to make my mortgage payments. I was eight years from retirement—that's what's left on my mortgage, eight-and-a-half years. But who's going to renew it now? I've probably got a hundred resumes out. It's tough at forty-seven. Nobody wants you at forty-seven." Is the mill viable? Yes, he says, it is. Should the taxpayers of the province shoulder a $300 million bailout? Yes, he says, they should—after all, the people in the Lower Mainland who now whine about putting tax dollars back into the north were never squeamish about the billions and billions they took out of the region.

Down on the fish docks, it's a similar story. In the heyday of the salmon and halibut fleets, Prince Rupert was the capital city of laundromats. There seemed to be one on every block—an illusion that was not surprising considering the frantic bustle at the height of fishing season. When the big Skeena runs came in, the place was transformed into a jostling madhouse of boats coming and going, nets drying, equipment repairs, expeditions to buy rations, to hit the bars—and to find a place to wash your stinking clothes. Today shops are boarded up and a curious lassitude hangs over the floats. From the morning coffee social in the eerie quiet that's settled on Pacific Net and Twine to Mayor Jack Mussalem's office at City Hall, an architectural treasure of creamy stucco, elegant art deco fluting and a stylized frieze of brightly-coloured Tsimshian figures, every conversation has a raw edge.

On rain-swept floats and in the frosty redolence of filletting sheds, from offices where the market has gone as soft as a mud flat to the noisy clatter of the United Fishermen and Allied Workers' Union hall, the questions are similar. Are we on course for the same disaster as the east coast fishery? Are far away urban governments, mesmerized by high-tech and silicon chips, preparing to sell out a way of life they neither understand nor care about? "When I was elected all these resource-based industries were healthy," admits the mayor. "Nobody foresaw 1997. Anxiety is the prevalent state now." And yet, there's that doppelgänger of optimism.

The mayor is enthusiastic about a joint project with the port that will see major refit of the idle Atlin Fish Plant, transforming it into a terminal for pocket cruise ships. These are smaller than the big cruise ships—even a mid-range cruise means 1,500 passengers—for which facilities are inadequate in most small coastal communities. "Pocket cruises carry 70 to 500 passengers," the mayor says. "We already have the infrastructure to handle that. Pocket cruises are attractive because they are more flexible and casual." Krusel concurs. This emerging trend provides a niche to which Prince Rupert can market its natural assets: proximity to Haida Gwaii, the restored aboriginal village of K'san just up the highway, the stunning landscapes of the Kitlope and the Khutzemateen, access to Ketchikan and St. Petersburg. "We have everything that any port in Alaska has, plus we have rail and road connections to the Interior. We are now on the cruise lines' radar screens. I think that in the near future we are going to have significant increases in cruise traffic. We are ideally located to expedite pocket cruises to the Alaska panhandle and the Queen Charlotte Islands. And our museum [in Prince Rupert] is world class."

Designed to evoke a longhouse, the new museum is a remarkable showcase of Tsimshian art that's equally comfortable hosting an exhibition of skateboard photography. The parallel displays seem natural and complementary. By force of history, Prince Rupert has always been a multicultural experiment. The city itself lies within a ring of villages dating back 10,000 years or more, all still home to powerful nations of the Tsimshian, the Gitksan, the Nisga'a and the Haida. Today, says Mike Tarr, almost half of Prince Rupert's resident population is aboriginal. To be sure, the usual problems of racial assumptions and ethnic superiority exist. Some people still refer to the strip between two beer parlours frequented by aboriginal people as "Apache Pass." But unlike other places at the interface of the two cultures, there's a powerful self-confidence among Native people who participate in the community—not from the position of supplicants, but from a position of relative power. "You don't find animosity when you know people and deal with them every day," says Gloria Rendell, chair of Prince Rupert's nine-person development commission. "The biggest thing that people [from the south] don't understand is that Native people here are not invisible, they are not a minority. Here, in many cases, they are the majority. They are highly visible. This makes for a very different dynamic." It's reflected in the development commission's reaction to aboriginal land claims. While southerners fret about the issue, Mike Tarr says its less a threat than an opportunity to develop new and exciting partnerships. "At first it seemed unfair but then, when you start thinking it through, it's what we have now that's unfair," he says. "The Natives aren't going away. People like me aren't going away. This could be a really good thing." For villages like Kitkatla, Metlakatla, Port Simpson, Kincolith, Skidegate and Hartley Bay, Prince Rupert is both service centre and cultural bazaar. It's a place to meet friends, exchange gossip, sell the catch, plot politics, cut business deals.

And the business deals are as important for non-Native Prince Rupert as they are for the adjacent reserve communities. Clayton Williams, who runs a construction company named Aggressive Concrete, frankly acknowledges that work provided by sympathetic local bands kept his company afloat. Census figures show that in 1996, the 3,015 residents of nearby reserves spent more than $9 million on consumables, goods and services in Prince Rupert. The average reserve family spends about $38,000 a year—and, contrary to popular mythology—pays an average of $7,698 a year in taxes. On average, more families on reserve own vacation homes than do families off reserve. Most interesting is the stereotype-exploding fact that while Prince Rupert exceeds the British Columbia average for families with computers, families on reserves exceed the Prince Rupert average. This, too, signals both opportunity and optimism.

Prince Rupert owns its own telephone company and it invested heavily in state-of-the-art high technology early in the game. "Now we have one of the highest rates of computer use in Canada," says Mike Tarr. "Microsoft and IBM are up here talking to our people at City Tel about how they do things. We know that the American military in Alaska needs a digital relay, we're looking at that—this would really plug us into the global network."

And yet, for all the promise of cyberspace, Prince Rupert remains rooted in the elementary forces of the natural world. Behind it, to the east, the Coast Range towers in a shining, almost impenetrable wall that seals off the immense wilderness of the interior. Before it, to the west, the steel-grey swells of the Pacific roll all the way from China. At the moss-covered ruin of the Union Steamship wharf the skipper in his bronze oilskins still stands in the rain, pointing out to sea: "We are out there." So is the future.

Yarrows Goes Down

I lived on Lime Bay just outside Esquimalt and on a do-nothing day I'd walk over to the dockyard by way of Work Point and Paradise Road, turn down Canteen Road, scuff through the bunch grass and smell the scent of wild poppies on the rocky outcrops mingling with the crisp scent of the sea. I was fresh out of high school, on my own for the first time, but just like a kid, I'd come to gawk at the grey warships, the gantries and cranes, the great echoing sheds and immense slipways of Yarrows Ltd. The shipyard was already in decline, limping from year to year on crumbs from the BC Ferries banquet. But to me it always seemed a way to glimpse the romance of the sea that I knew only from crumbling pulp novels salvaged from church rummage sale tables. And indeed, it was a window into that once rich past now peopled with fading shadows. Muscle and sinew of the country in war, a great pump of trans-Pacific commerce in peace, in its time Yarrows was touched by everything from fierce sea battles to the exotic silk trade. But times change and in a world increasingly characterized by high technology, Yarrows first became obsolete and finally redundant. When it announced in April of 1994 that it was closing the log on some of the most dramatic entries in Canada's stirring history of industrial labour, thousands came for the funeral.

I had to park way up the street at St. Paul's Church, consecrated as the Naval and Garrison Church in 1866. It's filled with memorials to ships and people lost at sea. Maybe they'll add one for Yarrows, foundered on the reefs of global change—too much old equipment and too many high-paying jobs to make it in a vicious new world where robots displace artisans and craftsmen. Like everything else around Yarrows, St. Paul's is a marvel of ingenuity. Built on the far side of Signal Hill, it was discovered that even the most devout prayer wouldn't spare the church from the fury of sea storms rolling down the Strait of Juan de Fuca from the open Pacific. So in 1904 they put the building on rollers and a steam donkey winched it over to its present sanctuary. I stopped a minute to admire the neat clapboard and stained glass in one bit of history before strolling down to the broken window panes, gaping doorways and peeling paint of another.

Yarrows was busier than it's been since its heyday in World War II, when five BC shipyards buzzed with 30,000 men and women building 244 freighters and sixty frigates, seventeen of those warships right in this shipyard. By 1970, all but one shipyard were gone from Canada's west coast.

At Yarrows' closing-out sale, the most extraordinary things were up for bids: dunnage and steel cables; metal presses; chains with links as long as your forearm; a 150-tonne crane; a 12-metre tugboat; a half-inch steel plate the size of your living room; self-contained fire-fighting gear. When I got there, Ritchie Brothers Auctioneers of Richmond had set up in the Module Shop, a metal shed the size of a football field. The auctioneer called for bids from a pickup truck, cruising by mounds of rubber hose, welding gear, cases of coated electrodes. Three days it took to sell it all.

I watched the grins and grimaces for a while, then headed down to Number 1 Ship Building Berth, a condemned cathedral of creosoted timbers. Chains clanked in the wind, pigeons cooed in the shadows and shafts of light lanced in through naked rafters. A century of honest sweat was salted into that wood. Eighty years ago, the ghosts were flesh and blood. They took only six weeks to repair HMS *Kent*, hit thirty-eight times by German shells in the Battle of the Falklands. In 1942, they converted the *Queen Elizabeth* from luxury liner to high-speed troop transport.

I tried to imagine the heady days when 700 ships a year jammed Victoria's docks. Sleek white Empresses, the Australians, the French line, fast silk carriers of the Blue Funnel Line—seventeen days from Yokohama to New York via SS *Protesilaus* and the Northern Pacific Railway—astounding in 1911. The Japanese battleship, *Idzumo*, came here for a refit under the Anglo-Japanese treaty that made us allies in 1914. Caught unprepared by submarine warfare and with no steel mills in BC, the shipyards responded to the desperation of 1917 by building freighters with wooden hulls. Tangible signs of those vanished workers were everywhere among the industrial relics piled so neatly for disposal. A welder's mask with voodoo eyes painted on it. Two perfect maple leaves cut from steel plate and shrouded in dust at the back of a locker. A cryptic note on a century-old wooden tool chest, the grime covering it so thick that the graffiti seemed embossed, unreadable except as braille. Glued above a tool-making machine, a photo of Paul G., 1952. In the sheet metal shop, a seniority list—Floyd Sidlick, June 11, 1964, down to Herman James, August 20, 1992. In one shop, a hastily scrawled reminder: Social Club Dance, Crystal Gardens. And in a corner, the other essential ingredient for a dance: girls' names and phone numbers jotted on the wall—Angie, Becky, Vickie, Margaret. Kim and Brenda. Karen. Judy.

The sky above Her Majesty's Canadian Dockyard was a landscape painter's perfect blue as I walked back through buildings stained by work and time. Across the harbour, nylon colours snapped at the mastheads. The breeze carried indistinct cries from naval ratings running signals up the flagstaff. Beside a 350-year-old fir log, no longer needed for service in the slipway, the wild poppies blazed with the brief, furious intensity of late spring. As I passed out the gate, I made a point of not looking back at the big sign that says "Yarrows is a good place to work."

ANYTHING TO SELL A PAPER

I heard about Mike the Dog from an old-timer sitting in the corner at the Fishermen's Lodge, one of those unrepentant, slightly out-at-the-elbows public houses that you can still find living incognito on the less-travelled side roads of Vancouver Island. You won't find the Fishermen's Lodge listed in the Yellow Pages under "restaurants" or "hotels" or even "beverage rooms." Full page ads for "Naughty Girls (Absolute Discretion Assured)?" You bet. But "Old-fashioned tavern with terry-cloth table covers, squeaky floors, no heavy metal band, draft on tap, real characters who've lived real lives and don't mind telling you about it and a stuffed steelhead that looks like a coelacanth"? No way.

Fishermen's Lodge nestles behind a levee on the north bank of the Oyster River about halfway between Courtenay and Campbell River. The levee is there because the Oyster has a habit of wandering, kind of like the palaver that goes on in the dark corners of the lodge. You never know who you might find in the Fishermen's Lodge. Maybe an unrepentant union organizer who'd rather talk about baseball than the great "pie strike" when the whole North Island walked off the job in a dispute over whether the bullcook had the raw power to cut off the faller who didn't return his dinner crockery promptly enough. (He did.) Or a prospector with placer claims on the river, hoping for just one nugget the size of a softball, although he'll confess that he'd probably settle for one the size of a baseball. Or maybe someone who'll tell you the sad story of Mike the Dog and the gravestone erected by his grateful master. From the get go it sounded like one of those jokes that gets embellished and burnished and traded around until it comes back ready to be pawned off on the gullible. Mike the Dog waited on table. Yup: he'd bring you a bottle of beer in his mouth, take your payment back to the bartender and bring back your change. Where? Well, where else would an honest dog sling beer but the Bowser Hotel?

No, wait. Bowser is a real place just north of Qualicum. It's even named after a dog of another sort, Battling Billy Bowser, still loathed and detested on the Island north of Ladysmith for sending the army in to train machine guns on strikers. The nasty little Tory hatchetman succeeded Glad-hand Dick McBride as premier and was soon so hated by his own party that its members ran ads in the next election begging the voters to drive "The Little Kaiser" from office. The voters happily complied, sending Bowser and the provincial Conservatives to the political wilderness they so justly deserved and in which they still wander. And there really was a Bowser Hotel, although it burned down decades ago. Faced with a story like that, what else was there to do? I went to Bowser to track down Mike the Dog.

Nobody at Bowser General Store had heard of the Bowser Hotel. The post office clerk suggested the Bowser Legion. There they knew everything about Battling Billy and about Bowser's Seventy-Twa—the 72nd Regiment—sent in to bust the miners when the Big Strike raged from Ladysmith to Cumberland from 1912 to 1914. They weren't so sure about the Bowser Hotel. "Hell, boy, that was forty years ago. That place was long gone when you were still supping on your mamma's milk." Some thought it had stood across from the Esso station. Others swore that had been the post office. The beer parlour was farther north. All warmly agreed it was this side of the Esquimalt and Nanaimo Railway tracks but not so far as the Legion itself. Had anybody heard of Mike the Dog? Nope. "Sounds like a put-on, son."

I squeezed through a rotting picket fence and the hole a helpful fat bear had thoughtfully made in the dense bramble thickets. A small mossy glade was hidden behind towering walls of blackberry vines that muffled the sound of the highway traffic. I found the ancient remnants of a bindlestiff's camp, rusted billy can still on the fire stones. A chicken coop, circa 1935, was filled to the brim with rusting cans and broken crockery. It sagged between a nice grove of stinging nettles and a bank of sweet-smelling garden roses gone wild among moss-covered apple trees. Next I found the scattered ruins of the post office. Further along was a peculiar slab of concrete poured around water-polished creek pebbles. There were scattered bricks from a fireplace and an old bar stool. But no sign of Mike's grave marker.

I turned to Ruth Masters, environmentalist before the word was invented, terror of all mouthy pro-development ideologues—"Share? Share the Stumps, they mean!"—and laureate of the local historians in Courtenay. Ruth suggested rooting through the files of the long defunct and almost forgotten *Comox Argus*. So I did. And there it was: January 9, 1941—Mike's obituary under the heading "Mike, Beer-serving Dog Dies."

> There is sorrow among men who take a glass of beer at the Bowser Beer Parlour for Mike, the famous beer-serving terrier who for the past eight years has helped his master tend the bar, is dead. He was hit by a car passing the tavern last Friday evening and died in hospital at Nanaimo shortly afterwards despite the valiant efforts of veterinarians to save his life.

> Mike was nine years old and had been trained by his master, Mr. (Charles) Winfield, to serve bottled beer to the customers, collect the money and return the change. He always kept a sharp lookout for empty bottles and would immediately return them to the bar.

When I wrote this story for the *Vancouver Sun*, a caller tartly informed my editor that I was either drunk as a thousand dollars when I wrote the column or the victim of a classic British Columbia tall tale on the order of elkhares, jackalopes and those pesky sidehill gougers—the critters that can only travel around the mountain in one direction because the two legs on one side are shorter than the other pair. This sceptic drew my editor's enthusiastic attention—editors are always enthusiastic when they are sure they've caught you in the act of being duped—to the BC Archives' Sound Heritage Series, Tall Tales of British Columbia, which records the story of Mike the Dog in a decidedly different light.

It seems one Ray Logie of Esquimalt called in with the story of Mike the Dog serving beer at the Bowser Beer Parlour during a Barrie Clark phone-in show on CKNW on July 14, 1982, and author Mike Taft collected the "tall tale" for posterity. But let me quote from the learned book:

> Now there was another one that was concocted by two old uncles of mine, while I was there, obviously for my benefit. At the time I was only twelve or thirteen. We were driving up-Island in a '26 Chrysler and we passed a little village called Bowser.
>
> There used to be an old pub there. I think it's burned down now. And clearly they were talking to each other, but with my rabbit ears, a little kid sitting in the car with them. It obviously was intended for me.
>
> As they passed Bowser one of my uncles said, "Oh, yeah, lookit, that was named for a dog: Bowser."
>
> "Yeah," and the other uncle said, "Yeah, they had that

dog in that pub and he used to serve tables and take orders." And the other uncle developed it.

"Yeah, it was there. And of course he could make change." And that was about the extent of it. But I remember the tall tale being developed by the two uncles. I don't know how long I believed that, but I finally realized I'd been had.

What about that obituary in the *Comox Argus*?

The caller had snorted. "That Ben Hughes'd write anything to sell a paper. Ain't he the guy that made up the story about the disappearing tribe on Forbidden Plateau?"

Well, maybe. And maybe not.

Ray Walker of Burnaby called to chide me for not going up to Cumberland to get the real dope on Mike the Dog. "That dog, it was quite a story when I was a kid. Every old-timer in Cumberland knows about that dog." Ray was born and raised in Cumberland. His dad, Bill Walker, ran the orchestra. Walker family roots run through the mines—three Walkers perished together in the 1901 explosion that killed sixty-four—although "I only worked in the tipple while I was going to school." He and his dad loved to fish—"I scattered my dad's ashes up there on the Oyster"—and they too would stop at Fishermen's Lodge, where I first heard about Mike. "The old man would give me a bottle of pop to keep me happy and send me outside. There used to be a deer at the Fishermen's Lodge after the Big Fire in '38. He used to chew on cigarette butts and drink beer from the big green ashtrays." I have corroboration on the Fishermen's Lodge deer, in any event. My wife Susan remembers it from her own childhood days playing outside while her dad went to whet his whistle after a long day of working the lower reaches of the river for sea-run cutthroat. Ray never saw Mike the Dog in person, but his dad told him the story and he certainly remembers the hotel from the big family outings to Qualicum Beach. "I'm certain there used to be a little grave for him there."

But I now have more than hearsay: I have an eye witness. Pat

Barnard of Tsawwassen remembers the Bowser Hotel well. Her dad, Fred Richardson, was born in Victoria in the 1890s. He used to stop at Bowser when he was driving up to fish the Oyster and visit with his sister.

"My Aunt used to own the Oyster Bay Inn, which was a restaurant. Her name was Dorothy Grant—maybe you've heard of Grant Logging on the Island.

"I was in that hotel four or five times. I must have been five or six and that would have been around 1938. The building, it had a long overhang-type porch and it was right up against the side of the road. There was a deer in a pen that I was allowed to pet.

"Gosh, the silly things that stick in a child's memory. The inside of the hotel was painted dark green and a kind of creamy beige in the beer parlour. It had a long, high counter for the bar. I remember the dog had to jump up on it. He was a smaller dog, the colour of a golden retriever. He'd bring the beer and he brought the change on a tray with a lip on it."

And with that I went back to the Fishermen's Lodge and bought a round for Battling Billy Bowser and the Bowser Hotel and for Mike the Dog, who served the beer and made change.

Fog

Normally, rosy-fingered dawn clambers over the snowy volcanic cone of Mount Baker and through the east-facing windows into my bedroom with boisterous enthusiasm. This morning I woke instead to a disquieting, almost reluctant light that reminded me of the bleached white of the moon snails we occasionally find half-buried in the sand at Shell Beach. I'm told bones from a long-forgotten battle between the Saanich and the Lummi people occasionally turn up along this shore. Perhaps that's just the misguided local folklore of newcomers, since the two groups are kin to one another. Maybe the bones are the aftermath of one of the epidemics that depopulated the coast, sweeping through a summer camp awaiting the salmon runs that swing around Boiling Reef and the end of East Point on Saturna Island. Whatever the source, however lost their origins might be in the uncertain mists of human memory, the bones are real, they do turn up. So do the coiled, sensuous shapes of ammonites buried in the ooze of some Cretaceous sea bottom 65 million years ago and sometimes found eroding out of the sandstone. For myself, I'm content to admire the lustrous, abstract shapes of broken cockle hinges and tide-polished oyster shells. And moon snails.

When I rose, Susan was brewing coffee in the galley kitchen and contemplating the merging of sea, sky and horizon into the same milky orb. A dense September fog had moved in overnight. I could barely make out *Dandelion*, as my daughter has named the battered yellow tinny inherited from her grandad that we keep moored just off the terraced sandstone that shelves into the tide. Around the house, the tops of fir trees simply vanished—along with my plans to steal away for a few hours fishing. No mail. No paper. This morning seemed a good time to sit back and enjoy the fog. We watched as it stirred on the ebb. Pennants snagged in an arbutus where a kingfisher ruffled his iridescent blue crown. The wispy streamers seemed to mimic the tree's fraying red bark.

We have only one word for fog. The synonyms mist, haze and vapour all deal with conditions that seem not at all the same—curious for the language of sea-going peoples who produced the likes of Drake and Raleigh. Science, on the other hand, defines eight kinds of fog. That thought put me in mind of Claude Adrien Helvetius. "Truth is a torch that gleams through the fog without dispelling it,'" he wrote in his 1758 preface to *De l'Esprit*.

Fog, according to science, is not a metaphor. Fog is simply a fact, a product of physical laws, an atmospheric state caused by condensation in the air at the earth's surface. Technically, fog is defined as a cloud in contact with the ground. This cloud has unusual properties, however. Fogs are colloidally stable. The tendency for some water droplets to grow rapidly by gobbling up adjacent droplets until they disrupt the equilibrium and fall as rain is not present in most fogs. Frontal fog accompanies rain that is warmer than the surrounding air. Steam fog is caused by water surfaces warmer than the overlying air. Upslope fog is caused by warm air flowing up rising ground to cold air. Isobaric fog is where air flows from higher to lower pressures. Isallobaric fog is caused by suddenly falling pressure. Radiation fog is where heat from an underlying surface affects cold air above. Advection fog is where warm air moves over a cold surface. Fogs may even consist of crystallized water. Glittering ice fogs form at low temperatures over areas of human habitation.

I went to my library to confirm the Helvetius quotation. There was a footnote. On first reading his book, Voltaire wrote back immediately: "Your book is dictated by the soundest reason. You had better get out of France as quickly as possible." Shortly thereafter, the state authorities burned it. With this gloomy thought in mind, I went for a walk in the fog, picking my way carefully among slick rocks and slippery drift logs, listening to the muffled acoustics of the foghorn speaking to the tankers and container ships threading the Gulf Islands down Haro Strait. By noon, the sun shone again on my daughter's beloved boat. Helvetius, I concluded, was right. Fog obscures the truth only temporarily.

For the great American poet Carl Sandburg, the fog came "on little cat feet" and sat looking over the Chicago waterfront on silent haunches. For T.S. Eliot, it was the thick, yellow London fog "that rubbed its back upon the windowpanes" like a burly marmalade tom. But in the great expanse of western Canada, where the fog can swallow Edmonton's gleaming skyscrapers or submerge the Lions Gate Bridge, where planes fly into it and never come out, where ships have been known to sail into it and then off the edge of the earth, these pretty urban images give way to something deeper and more disturbing.

In British Columbia, perhaps it's the presence of the sea and our collective memory that out there on the Graveyard Coast, the fog is the silence that leaves us listening as the sails flap, straining for the too-late sound of combers on Wreck Beach or the suck and thunder of surf against the Broken Group. In Alberta, perhaps it's the dripping gloom of ancient mountain valleys cloaked in moss, where footfalls make no sound and the dark peaks suddenly vanish into a coiling, white canopy that erases the trail and hides the reference points that show the way back down. Or maybe it's just the way it occasionally boils out of plowed Saskatchewan farm fields like spirits come to haunt the low ground, hiding the sloughs and the gleaming shanks of newly bare aspen stands alike.

In any event, when the season of fogs arrives again each fall to tell us the harvest is just about over, we westerners don't think of Sandburg's delicate kittens. We think of the steamer *Pacific*, lost with all 250 people off Cape Flattery, or maybe of Dancing Annie, the phantom said to swirl in her raiments of mist on Comox Hill, dancing away—from what? Some say the smallpox, for the fog was once believed to carry contagion. Some say just another brutal husband. "Ain't seen her myself," old-timers in the Comox Valley say. "But, hey, I know a guy..."

Or we think of the austere industrial beauty of a steel seiner, nets piled high, ghosting out of a fog bank off Sointula, huge and startling along the awestruck ferry rail. Or of strange optics along the horizon. South from the Sand Heads, that vast, treacherous, alluvial fan that spreads just beneath the waters off the Fraser River's mouth, lenses of mist will bend the light and lift supertankers from below the curve of the earth into the shimmering sky. From the northern inlets we might remember the sight of a Beaver float plane flying low, mountain walls on either side rising into ragged murk, everything timeless and immovable except for that bright and agile yellow bead dodging the white clag and silver squall lines over black, black water. Or maybe we just hear it, the way it distorts sound, muffling everything and then suddenly transmitting with perfect clarity the clink of a bronze rowlock, the gabble of Canada geese in the barley stubble or an intimate conversation from the other side of the field between people we can't see.

From tundra airstrips to prairie wheat fields, from mountain lookouts to towboat decks, the weather reports across the West are alive with fog warnings in fall, a reminder that our world is once again tilting into darkness and away from the sun and that, for all its ability to evoke the sensibilities of poets and painters, this ephemeral, insubstantial coalescence of water vapour has an enormous economic dimension in the here and now.

The cost of obscured visibility is difficult to measure, but it's safe to say that it's in the billions. In one year alone aircraft groundings in the United States cost $75 million. It was in a fog that a river barge struck a bridge, displaced the tracks above and caused the worst train disaster in American history when the locomotive and its passenger cars simply sailed into the deep, murky river. For all our high technology, fog still causes widespread disruption of transportation systems for several days out of each year in most temperate countries. It can foil the most sophisticated electronic weapons of the superpowers—as the pilots of war planes discovered, day after day, during their air attacks on Serbian army targets in the Balkans during the war over Kosovo in 1999. There the fog of war was more than metaphor. Instruments and computerized navigation may mean pilots can now land in zero visibility, but flight rules say that if visibility on the runway drops below 600 metres, a landing can't be attempted. And pilots are not supposed to depart if their destination is socked in. So fog still closes airports as it did for Roaring-Twenties bush pilot Wop May, even in a day when global positioning technology can place an aircraft within centimetres. In addition to airports, fog closes freeways and sea lanes. It's been responsible for catastrophic ship wrecks, train wrecks, highway pile-ups. Mingled with exhaust fumes and industrial gases it can pose an enormous health hazard.

On December 5, 1952, a strong temperature inversion settled a dense fog of the kind they call a "pea-souper" over London, England, trapping residential smoke, factory fumes and auto emissions. The fog turned yellow, then brown. Four thousand people died of "respiratory complications." A less euphemistic term might be "asphyxiated." Tens of thousands more had their lungs permanently damaged.

Today, half a century later, public health authorities voice similar concerns over the air quality in the Fraser Valley as the incidence of asthma and other respiratory ailments parallels the rise of air pollution, spiking sharply upward with temperature inversions and relative humidity. If Sir Francis Drake complained of the "vile and stinkinge fogges" that drove him off the west coast during his bold circumnavigation of the globe in 1578, he might forgive us for saying that he didn't know what he was talking about. The industrial age's brew of toxic gas and suspended water vapour—smog and fog—is medically associated with elevated levels of bronchitis, emphysema, asthma and lung cancer in exposed populations. In some cases, points out science writer Joel Myers, fog and air pollution co-exist in a lethal symbiosis, the haze of industrial particles and exhaust emissions helping to generate a fog and then causing it to resist dissipating. The fog, in turn, reflects back solar radiation, providing a favourable environment for the further accumulation of air-borne pollutants.

For most of us, fog is simply cloud that has descended to the ground. Like every cloud, it's comprised of tiny water droplets forming around dust motes in the atmosphere. Technically, such a cloud is not a fog until it becomes sufficiently dense that horizontal visibility is lowered to less than 1,000 metres. If you can see farther than that, it's not a fog, it's a mist. Whether you take your pleasure, as Malcolm Lowry did, in watching a rugby game played in the ground mist at Brockton Oval, or you share Emily Carr's Zen-like appreciation of fog-draped cedars, there's an ethereal beauty inherent in the combined altering of sight lines and sound and tactile sensations.

Even the physics of fog is an elegant, complex marvel. All fog is formed in the same basic way, by water vapour condensing on microscopic particles of dust, pollen and smoke that are always present in the air. This process begins whenever the relative humidity of the air exceeds saturation by a fraction of one percent. Relative humidity—the ratio of water vapour present to the amount required to saturate the air—is a function of three processes: the cooling that takes place when air changes in density by expanding; the collision of two humid airstreams with different levels of humidity and different temperatures; the direct cooling of air by radiation.

If the mechanisms seem straightforward, the results are diverse. There seem to be as many fogs as there are changes in the weather conditions. In the high Arctic, there is ice fog, a glittering cloud of tiny frozen particles that tends to form in extremely cold conditions and can bend incoming light, creating a hazard for pilots attempting to land in the snowy white landscape. It's been found, for example, that visibility in an ice fog is considerably worse than it is in a fog containing the same concentration of water droplets. Around settlements, water vapour escapes from chimneys and is released by human respiration. Speak and your words sparkle. Where enough crystallizes in the bone-dry air, we call it a habitation fog. Steam fogs occur when cold air moves over a warm, wet surface and becomes saturated with evaporating moisture from below. Thermal convection carries the fog upward as it forms and streamers of it appear to dance like ghostly Annie above the surface of the water. This is a phenomenon frequently observed on the prairies in autumn because with clear skies and long nights, the land is cooling rapidly, while the water in lakes, rivers and sloughs still contains latent summer heat. Farther north a similar thing occurs when currents or winds create sudden openings in the pack ice. Because the differences in temperature and relative humidity are so much greater, the fog generated is much more dense and is known as Arctic sea smoke.

Along the ragged BC coast, with its deep fiords and thousands of islands, we tend to experience advection fogs, formed when warm, moist air slides gradually over a wet, colder surface. This is a frequent occurrence when cold and warm ocean currents become close to one another, or when currents run close to a colder shoreline. As a result BC, with its pattern of upwelling from deep ocean trenches and the Alaska and California currents off shore, has more days of more fog cover—226 days a year—than any other region of Canada. Newfoundland, where the Gulf Stream passes through the ice-laden seas of the North Atlantic, ranks as the second foggiest place (206 days a year). The least foggy part of the country is Saskatchewan (37 days a year) where there are no significant bodies of water or changes in elevation to trigger the temperature exchanges that cause cloud formation.

During the autumn months, those who live inland and "on the continent" beyond the damp and temperate coastal zones experience radiation fogs. They form over land on calm, clear nights when heat radiates out of the earth into space. As the temperature of the ground falls, air close to it is chilled below the dew-point and water vapour condenses above low points. This fog then substitutes for the ground as the surface that is cooled by radiation and then grows upward in altitude, increasing in density as long as there is enough moisture in the air above. Finally, those who live in the Lower Mainland are susceptible to frontal fogs which form when rain falling out of warm air above a frontal surface evaporates into a layer of colder air below until it is saturated. With cold outflows from the Interior following the Fraser River and moisture-laden warm fronts arriving frequently from the Pacific, we often get frontal fogs hanging just below the mountaintops which most of us mistake for low-lying cloud.

Science is just now beginning to understand that while these fogs can be inconvenient to travellers and to commerce, they also have a vital role to play in the ecosystems where they occur. On the coast, fogs hold and transport huge volumes of water both to the temperate rain forests of the coastline and to what are now called cloud forests at higher elevations. Condensation waters fragile grassy slopes without eroding them. Some scientists even suggest that fog droplets might be captured in desert areas in large artificial fog collectors, providing a low-tech source of water for agriculture and forestry.

When all the science and technology is said and done, fog in its ambiguous form is most important to us as individuals. It cloaks the familiar and distances us from the tangible. It reintroduces mystery to a world predicated on certainty. It evokes again that ancient mythology in which a wall of mist separated us from that other world, the unpredictable world of the unknown, a world in which anything could happen—even the ghost of Dancing Annie.

Tin Hats and Siwash Sweaters

It all started with the tin hat. The hat's not really made of tin, of course, just the same waxed cotton canvas under which generations of loggers have sought shelter from these endless West Coast rains. Half the time the rain is somewhere between a drizzle and a fog, a kind of heavy dew condensing on everything. The way it drips off the canopy it might as well be rain. It runs down your neck and under your flannel shirt and soaks your Queen Charlotte's tuxedo—which is what some folks around here still call their one-piece Stanfields. A proper brim on a good tin hat spills the rain past the Grand Canyon of your collar.

My dad wore a hat like it almost fifty years ago on the west coast of Vancouver Island when he was laying down spruce as a gyppo contractor for Muir Brothers sawmill. His show was so haywire—and so flat busted—that when the big two-man German chainsaw broke and he couldn't afford to back order the part to fix it, he handlogged the rest of his lease to fill the contract. He filled it on time, too, even though he had to peel the spruce logs with a flat spade. But hey, a deal's a deal. Once upon a time in this country a man's word was as good as his handshake.

And my father-in-law, who preceded me through the *Vancouver Sun*'s newsroom sixty years ago, wore a tin hat when he worked as whistlepunk and sparkchaser for Butcher Bloedel—as Bloedel, Stewart & Welch, the forerunner of MacMillan Bloedel was once rudely called. That was before taking up newspapering so he could get married, live in town and eat a dry lunch. Whenever I drive the Island Highway now, we go through the ritual of slowing for my daughter as we pass the BC Forest Museum and offering a toot to old "One Spot," the 45-ton wood-burning Shay that once hauled the logs her grandfather helped cut.

So when this year's Christmas present turned out to be a genuine tin hat, I was pleased. Even after thirty years it takes a woman with the right stuff to know what lurks in the heart of a man. You can keep your silk ties. I'll take a tin hat every time. No true coast rat should be without one.

My hat is a Filson, purveyor of quality headgear to people who have worked outdoors since the skids on skid road referred to bunkhouse foundations rather than peoples' misfortunes. And let's be clear, that skid road was East Cordova in Vancouver, not Yesler Street in Seattle. Our pals to the south take credit for coining the term skid road just because Henry Yesler built one near Pioneer Square. Actually, for those who care to be accurate about such things, skid road has its etymological roots in the pre-revolutionary Royal Navy dockyards.

Anyhow, I tried on my new chapeau. But the truth is, nothing looks so forlorn as a tin hat and one of those uptown Gore-Tex squall jackets from Mountain Co-op or Eddy Bauer.

"For that jacket you need aviator shades and a cranberry baseball hat," I was informed. "For that hat you need to ditch the Gore-Tex and get a siwash sweater."

Now, before the guardians of politically correct speech take umbrage at the word siwash, let me remind everyone that until it fell into recent disfavour thanks to misuse by the stupid, it was a self-referential term in Chinook, the vigorous trade jargon that exemplifies the cross-fertilizing fusion of cultures that characterizes the best of this coast. My white-collar, high-ratio mortgaged colleagues from the east are difficult to convince, but a Chinook dictionary was once as essential as *CP Caps and Spelling* for reporters venturing beyond the city limits. My own tattered copy was bequeathed by a predecessor who departed our vale of tears so long ago nobody remembers he was here. The dictionary was dog-eared long before it got to him. I thought of what he would say if he ran into me wearing my tin hat with my Gore-Tex jacket: "Hyas pelton, mika." Right: pretty silly.

Nope, a quality tin hat demands a quality sweater. And not one of those trim, trendy, touristy Gastown models, all chocolate-coloured reindeer leaping across snowy wool. A machine-knit imitation won't do, either. A tin hat demands a sweater that is knit from bulky hand-spun wool, the kind of sweater that will keep you warm even when you've had a good soaking. It has to be sloppily oversized and knit from coarse, lanolin-rich, rain-shedding yarn. The pattern should be a traditional grey snowflake on dirty white. And it has to be made by the right woman.

Where else to begin on a sodden winter weekend with the rivers in spate than with a spur-of-the-moment expedition to the Cowichan Valley on Vancouver Island's east coast? My wife is the guru here. Susan's family connections to the Cowichan Indian Reserve reach back almost to the First World War. She passed on to me the story of the "ool church," paid for dollar-by-dollar with shepherds' and knitters' contributions. Her practical advice: stick my nose in first at Hill's Koksilah Store and see what's on the rack.

Hill's is more than a commercial venture, it's also a rich piece of local history, one of those welcomed alternatives that helped aboriginal artists find prominence in mainstream markets outside museum collections and gift shops. Indian bands run competing retail outlets now and gone are the days when a Cowichan sweater went for $15 at the sporting goods store where you chose your spoons and Lucky Louie plugs. Old-timers liked their sweaters long. Fishing from dugouts for winter springs in Cowichan Bay, they would pull them down and squat inside a kind of tent.

Originally, Cowichan women wove cedar bark, dog hair and mountain goat wool. When Scottish crofters came as pioneers about 150 years ago, they brought sheep and knitting needles. Traditions merged. With needles fashioned from whalebone and golden ninebark—later from telephone wire and chopsticks, today from plastic—Cowichan knitters began producing unique waterproof sweaters perfectly adapted to the coastal climate. Traditional designs ornamented sweaters prized at first by loggers, timber cruisers, hunters and deckhands on the double-enders. Today they are recognized worldwide as art.

The art went uptown, as it will, and so did Hill's, which is why most of us associate the name with that trendy gallery on Water Street in Vancouver's Gastown. But the original Koksilah store is still there, a creaky frame house with gables, an old fashioned veranda and a floor that squeaks. We arrived exactly five minutes before closing on New Year's Eve (hey, I said it was spur of the moment). Colleen McCartney—who had just become Colleen McKay—really wanted to close up. Colleen was a remarkably good sport for a woman about to leave on her honeymoon. A middle-aged guy fussing over patterns, colours and size was definitely not high on her agenda. But she let me in to look anyway. There was nothing that fit that I liked and nothing I liked that fit. But Colleen had a New Year's gift for me: a name.

Gloria Tommy has been knitting Cowichan sweaters for more than fifty years. She learned at the knee of her mother, Susan Harris, who knitted for more than sixty years and learned when she was "little, little" from her mother by lantern light. Gloria's daughter Valerie knits. Granddaughters Ruby, 11, and Cecily, 8, both knit. Even her grandson David knits toques and mittens when he needs pocket money. The matriarch of this sweater-knitting dynasty—twenty-four grandchildren and five great-grandchildren—agreed to make me a sweater, properly overlength, snowflake pattern, same style as graced the backs of Dief the Chief, Bing Crosby and Prince Philip.

It took 2.2 kilos of carded wool. She tweaked it and on her own spinning wheel spun it into characteristic Cowichan yarn—one that's noted for a unique twist that gives it surprising elasticity. She used three balls of yarn the size of basketballs to knit my sweater—black and grey for the patterns, white for the background. When I collected it, we went for coffee. I told her a story about my father-in-law and a dugout canoe. She told me her father-in-law carved those dugouts, maybe even that one. Then Gloria told me quietly that she'd added a traditional wave pattern to frame my snowflakes—I hadn't asked for it, but the design wanted it. She was absolutely right. It was perfect.

What I did next was another West Coast tradition. I put on my siwash sweater and my tin hat and I went for a long walk by the river in the rain. It came down buckets, dimpling the pools where the trout were resting and hissing through the trees. When I got back, my high-tech boots with the Gore-Tex liners felt kind of soggy. The rest of me was warm and dry.

HARVEST

The earliest accounts dismissed British Columbia as a barren, desolate wilderness, rendered unfit for farming by a harsh climate and its stark topography of naked rock and arid cordillera. With its gloomy coastal rain forests and only the thinnest mantle of acid woodland soil, with its immense snowfields and expanses of ice-scoured alpine, much of it so high in altitude that it is less hospitable to plants than the Arctic tundra, even today barely two percent of BC is actually under cultivation—and what little still remains to be farmed must be considered marginal at best. Yet in scarcely 150 years of enterprise, what pockets of fertile soil are to be found have been fashioned into some of the most productive farmland on the planet. And BC now has the distinction of being the only province in Canada where family farms are actually on the increase. Numbers grew by 14 percent over a five-year period. Notwithstanding the weather-related disasters that dog recent harvests—rain across the wheat fields, hail in the fruit belt, potato blight and water-logged fields across the Lower Mainland and Vancouver Island—agriculture in the province remains an astonishing success story.

From the oil on the salad to the glaze on the ham, from the wine in the glass to the kiwi flan, the astonishing cornucopia that is British Columbia's autumn harvest supplies virtually everything that adorns the urban Thanksgiving dinner table. Twenty-four thousand tonnes of cranberries will leave BC for North American tables in the fall. They will garnish $27 million worth of turkey dinners and $158 million worth of chicken. And from the 7,250 hectares farmed as market gardens would normally come $156 million worth of fresh vegetables, although losses to root vegetables might top $20 million.

BC's first hardscrabble venture into agribusiness was established by the Puget's Sound Agricultural Company at Esquimalt in 1852 to fill Hudson's Bay Company contracts with Russian settlements on the north coast. Who back then imagined that Fort Okanogan and its fur brigades would vanish utterly while the narrow benches and sagebrush deserts left by dwindling ice age lakes would blossom into Canada's most important apple-producing region? Or that the richest gold strike in the interior uplands would not be a panful of nuggets but the discovery by Joseph Guichon and John Douglas of wild bunch grass capable of sustaining almost 300,000 beef cattle from the Nicola Valley to the Cariboo? Or that the 1,200-kilometre cleft carved in the mountains by the Fraser and its tributaries, the glaciated offshore islands and the silty alluvial fans of coastal estuaries would pour forth the $2.2 billion torrent of cereals, vegetables, fruits, dairy products and meats that we will formally celebrate on Thanksgiving? Or that the brooding forests inland from Fort Victoria would give way to rolling dairy farms, wheat fields and market gardens?

In BC, 97 percent of farms are family businesses. In a sense, Thanksgiving is also celebration of the ingenuity, tenacity, wisdom and entrepreneurial skill of the 21,575 farm families who endure and even prosper in the face of market uncertainties and daunting external risks that many of us find unfathomable. "The season of mellow fruitfulness," John Keats called the harvest season, invoking an English country gentleman's image of how the farm year should come to its satisfying conclusion. But here in the westernmost province, Thanksgiving is no more than a brief respite from the gruelling race against implacable time and capricious weather that takes place across a complicated sequence of the eight regional micro-climates which punctuate a territory the size of western Europe.

For Dawson Creek wheat farmer Frank Breault, it's a capital-intensive, highly automated, high technology venture. He, wife Dona, son Greg and hired hand Darren Thompson need $500,000 in equipment to harvest fields which cover eight square kilometres. For Penticton fruit grower Lucio Almeida, it's a labour-intensive but scientifically precise excercise in genetics, chemistry, timing and human resources management that enables him to pull around 200,000 kilos of choice specialty apples from five hectares of orchard located only 17 metres from an 88-unit suburban town-house development. For Keremeos vintner Joe Ritlop, it's a low-tech but surprisingly diversified family operation on nine hectares of grape vines next to the Similkameen River. His vines yield twenty-one different wine labels that are largely snapped up in word-of-mouth and drive-by sales at the farm gate. For Richmond vegetable farmer Harry Hogler, it's not so much getting things to grow in the rich silt of the Fraser River delta as it is the battle against constant urban encroachment, an ability to read volatile commodity markets where pricing changes by the hour, and developing the innovative retailing strategies that permit him to compete with transnational corporations. For legendary Victoria apiarist Babe Warren, it means the remarkable logistics of setting and servicing bees in places as remote as the Carmanah and Walbran valleys and then gathering her world-renowned fireweed and wildflower honey from 3,200 hives scattered across the 31,285 square kilometres of Vancouver Island.

And yet for all the graces said and thanks given and fellowship shared on Thanksgiving, few of us will really take much time to think beyond the groaning market stalls and well-stocked supermarket coolers to consider the sweat and daring behind this rich harvest in all its magnitude and diversity. Perhaps this is because we still tend to define ourselves with the muscular industrial archetypes of timber and mines and moving freight, or the notion that we're arriviste white-collar financiers in glass towers who are now above all that toiling in the dirt. The attitude is reflected in the provincial budget, which sets out the indisputable strategic priorities of the province. Since the New Democratic Party took power, the provincial agriculture department's budget has repeatedly suffered the most draconian cuts—down 57 percent from $115 million to $48 million.

Yet how many British Columbians, mesmerized by Asian investors and Washington lumber politics and the bailing out of aging pulp mills, are aware that 200,000 jobs here rely on agribusiness. Or that while logging jobs shrank by almost half, jobs in the agricultural sector were exploding with 41-percent growth? How many perceive their province as one which has a smaller percentage of arable land available than any province save Newfoundland, yet which equals the agricultural output of Manitoba, where three-and-a-half times as much farmland is under cultivation?

So perhaps there's a small irony to be found in the fact that the economic miracle BC celebrates with the Thanksgiving holiday really begins on the far side of the Rockies. As far away from Vancouver as is California, BC's northeastern prairie spills into the vastness of the Great Plains where 85 percent of Canada's farmland is found and which descend unhindered all the way to the Gulf of Mexico. If BC sometimes believes itself alienated from Canada, the Peace River feels itself no less distant from Victoria—and as neglected—as the Lower Mainland claims to be from Ottawa. "Over $1 billion goes out of this region every year in taxes and royalties," says gritty Dawson Creek Mayor Blair Lekstrom. "Only $4 million comes back in—you do the math." Even the geography of the Peace is incongruent with the rest of BC. In a province where all sensibilities seem finely tuned to the west and the south—the same directions in which the rivers run—the rivers of the Peace all run east and north.

The natural orientation in Dawson Creek is not Vancouver and Pacific Centre but Edmonton and West Edmonton Mall. And retailers frankly acknowledge that one burden of doing business is the fact that a few kilometres drive to the east there's no provincial sales tax. No big deal for sundries, but significant when you're a young family looking for a new fridge or kitchen range.

In a province where mountains crowd both the sky and public consciousness, this is a landscape that blazes with the vivid patchwork of yellow canola, blue-green rye and golden barley: where towns carry names like Bluesky and Sunset Prairie and the vault of heaven sweeps from horizon to immense horizon.

Under that velvet expanse, with stars glittering in the west and the first silver splinter of dawn wedged into the eastern rim of the world, you'll find Frank Breault preparing for a workday he hopes will run to nightfall and beyond. Breault seeds wheat, rye, canola, barley, forage peas, alfalfa and hay into three-and-a-half sections pre-empted by his father and his maternal grandfather eighty years ago. He's proud of the fact that he still farms the first quarter section, with its three-and-a-half by five-metre homestead of mud-chinked, hand-squared logs. He's even prouder of his heritage. He's what Quebecers like to call *la laine vrai*—the pure wool—the direct descendant of farmers who settled New France in 1663, a living example of those invisible bloodlines that knit a vast country together beneath the intemperate rhetoric of politicians.

For him the harvest window is narrow—only ninety-seven frost-free days—and fraught with the risk of early snow. In the region were 85 percent of BC's grain is grown, every hour spent taking off the crop is an hour secure in the bank and free from worry. The evenings close in early in these northern latitudes, but the farmers adapt. Their machinery is equipped to illuminate the fields with the brilliance of a movie studio. "Oh yeah, there are times when guys go fifty hours without stopping," says Bill Greenhalgh, grain program coordinator for the provincial government. "They just pile the grain at the side of the field and keep going."

Frank begins his day by running his thumbnail down a kernel of wheat to see whether it holds the indentation that signals starch and moisture content. Then he scuffs the stubble to see if his boot tips get wet from dew. If the crop is ready and the straw is dry, he'll climb into the computerized, air-conditioned cockpit of his big, green $250,000 John Deere combine and start taking up the grain. Then, flanked by outriding hawks that stoop to the field mice scurrying from the windrows, serenaded by Canada geese mustering for left-overs in preparation for the long flight south, he'll keep at it in a plume of dust and chaff until the night dew settles. Breault, Darren the hired hand and son Greg will spell each other at the controls so that the combining is continuous. Sometimes one will drive a 10-ton truck in parallel and the wheat will be loaded even as combining goes on non-stop. The other will bale hay for export to the south, the conversion of hayfields in the Thompson-Okanagan to ginseng farms having created a whole new domestic forage market.

Harvest on a big spread today involves conferencing by cell phone, fax machines and computer link-ups. BC farmers lead the country in personal computer use—an increase of 87.4 percent over 1995. Looking into the Breault's combine it's easy to see why. The array of of instruments is as complex as those in an aircraft cockpit. They inform the operator of everything from system diagnostics to how many hectares have been harvested. "If your harvest capacity is correct," Breault says, "you should be able to get off five percent of your harvest per day. So most of us figure that three weeks of good weather will do it." In the late 1990s, poor weather resulted in 50 to 60 percent of crops being caught in the fields, which is why it's not uncommon in the Peace to see combines two and three abreast, floodlights illuminating the fields, taking the crops off until the last possible minute.

Lucio Almeida's harvest is a phased, orderly affair, dictated by the timetable to which different varieties of apple mature. "We start with Galas," he says. "Next we get off the Golden Delicious. Right now Gala is the hottest apple and it brings the most money." In fact, a prime Gala apple—a yellowish variety with rosy streaks—brings more than four times the value of a plain Golden Delicious. Fuji, a crisp, tart Japanese variety, is another hot seller, bringing only a mite less in return than a Gala. "But Fuji is risky," Almeida points out. "It's a very late apple. You can get caught. Four years ago we had a cold snap—it went down to minus 10 in early November and it caught everyone with their pants down." It's one of the reasons, he says, that he farms on one of the sloping benches above Skaha Lake, just below the home he says NHL goal tender Andy Moog—a Penticton native whose dad played for the world champion Vees—has bought. On the slope there is less chance of still air and a killing frost.

The rigours of the weather notwithstanding, a few weeks earlier he'd stood at the edge of his orchard and watched a sudden hail storm wipe out his neighbour. Long-range strategic planning is the foundation of fruit farming. In an age of instant trends and volatile tastes, will the Fuji and the Gala hold their market value? This is a critical question. It costs Almeida close to $100,000 a hectare to replant and when he does it takes five years for the new trees to really produce. "Fuji and Gala hold their price right now," he says, "because there is limited supply. But we know that the Americans down in Washington are replanting these varieties—so how do we plan?"

One strategy is to become increasingly efficient.

"The old trees planted at the turn of the century take a 16-foot ladder to pick. You can't afford that kind of time anymore. You get about 112 of those old trees to the acre. If you go to dwarf hybrids you can pick from the ground and plant 1,000 to the acre—I know one guy who has 5,000 to the acre." Mind you, he laughs, the reason pioneer fruit farmers grew tall trees became clear when the deer moved through his new plantations and ate every bit of fruit from the dwarf trees. Not to mention the 240-kilogram black bear that grew so fat on apples it wouldn't fit into the conservation officer's trap. Now his orchard is surrounded by a five-metre fence.

Another strategy he employs is production of high-quality fruit that is tailored for specific fresh market niches. "Fuji and Gala, the bigger the apple the better," he says. "Spartans—I want a smaller apple. Why? It's a winter apple and moms don't want a big apple to put in their kids' lunch boxes. To the moms who make the buy, a big apple means a wasted apple." Where many orchardists abandoned Red Delicious production a few years ago, he says, he stayed in because he anticipated consistency in export demand. Today choice Red Delicious remain one of BC's best selling exports, particularly to Asia and Britain.

The most important element in producing for quality retail markets, Almeida says, is becoming intensely focussed on the dignity and value of human resources. He pays his pickers by the hour instead of by the bin because he wants them to feel comfortable going more slowly. He pays more than the going rate because that way his workers will take care of him by taking care of his fruit. He hires women, although they have less physical strength, because he says they're more willing to learn new ways and they're more interested in team outcomes than individual performance. "A lot of growers, they work toward this [harvest] all year and then they try to cut costs on the labour. That's foolish. If a person does the job and does it the way I want, I don't mind paying more. If you try to cut costs now you can wind up with 30 per culls on your Golden Delicious or your tender Macs." Almeida must be doing something right. Surinder Khaira and Parmjit Dhaliwal come all the way from Surrey to pick for him. Charanjit Garcha has returned every summer for eighteen years.

How does Almeida feel about the future of fruit farming given free trade and the prospect of Washington orchardists dumping into the BC retail market? "We are optimistic. We farm for the future. How can you not be optimistic when your orchard takes so many years to mature?"

Joe Ritlop is another optimist who believes small is beautiful and that as baby boomers mature the market increasingly demands specialized quality over cheaper mass production. Under the huge K formed by talus chutes on the sun-blasted face of Keremeos Mountain, Ritlop's St. Laszlo Estates vineyard is small but diverse and intensively managed. It's an oasis of coolness in the scorched landscape, the vines drooping their plump clusters of Chardonnay and Pinot Blanc and half a dozen other varietals into the jade-green shade. These grapes might serve as a symbol of the agricultural sector's amazing ability to adapt to changing conditions. Ten years ago, challenged by the increasing sophistication of the consumer's palate, producers were forced to address the demand for quality by tearing up vines which produced inferior grapes. Between 1981 and 1991 the number of hectares growing grapes fell by half. In the late '90s the sector is back to about 70 percent of its original strength. But average annual sales from the 833 hectares in BC vineyards have doubled—a remarkable productivity gain.

The initiative is evident in Ritlop's vineyard. At each end of each row of vines is a walnut tree, eighty in all, producing a crop whose retail value he estimates at around $1,500 a tonne. And shielding the vineyard from Highway 3 are dense thickets of recently planted hazelnuts—now emerging as a $400,000-a-year specialty crop across the province. "We started growing grapes here in 1978," Ritlop says. "We started with table varieties and then got a retail licence and moved into wine in 1985. "A lot of egomaniacs start out big and then they become a statistic. Here it's like a kibbutz—it's all family, eh. It's really labour intensive. Everything hands-on. You can't automate a vineyard. I'd say I put in ten times the labour of one of those prairie grain farmers. And we do most of it ourselves." Behind Ritlop, attended by squadrons of drunken wasps, the grape press is working overtime and the juice—250 litres for every 500 kilograms of grapes—slurps into big plastic pails, the first step in the processing of close to 50,000 litres. His wines range from dry reds to sweet central European dessert wines and ice wines—produced by leaving late grapes to freeze on the vines—and he says they pretty well sell themselves.

For him Thanksgiving is just a break in the harvest. "I won't be finished now until the end of October. We had a poor spring but then things seemed to catch up. The only problem I have now is with my little feathered communists: the starlings think all property should be held in common, especially my grapes. And frost, don't say that word around here. It's scary."

At Richmond Country Farms, which supplies its bustling retail market at the intersection of Steveston Highway and Highway 17 from just one square kilometre of rich flood plain, the challenge is finding ways to compete with giant supermarket chains. Norm Farrell, one of Hogler's managers, thinks one way is by helping urban kids to reconnect with the agricultural community that sustains them, which is why the farm puts about 30,000 Greater Vancouver school kids through its annual fall Pumpkin Patch. "We grow 30,000 pumpkins just for this. They get to go out in the field and select their own. No kid leaves without a pumpkin." Just how alienated are these city kids from farm life? You wouldn't believe it. I mean, we've been asked by parents 'Can't you cover up your fields so the kids won't get wet if it's raining?' Sorry, mom, but this is a working farm."

Aside from kids, Farrell would like to educate urban municipal politicians about the realities of farm business. Planners don't think twice about redesigning traffic flows to provide protected left turns into shopping malls, he says, but when the adjacent road was improved to serve a ten-screen movie complex looming just up the way, westbound access to the market was made illegal. "We've lost a huge amount of retail volume," says Farrell. "In terms of lost profitability our loss is going to be in six figures. We're being pushed by the public to stay on as a farm but other forces push us the opposite direction. Yet I believe the value of an urban farm like this is really high. I don't want to sound like we're whining. When you are in here in October and it's full of all those kids it's just heartwarming." The municipality's response to the popularity of the event, he says, was to post no parking signs on all the road shoulders.

At Babe's Honey Farm in the rolling countryside of Vancouver Island's lush Saanich Peninsula, Babe Warren is putting up the flow from an operation that accounts for 10 percent of BC's hives. At seventy-nine, Warren has been raising bees for fifty-two years, processing it at her rural plant and retailing honey prized like none other. "We built our clientele on honey that won't drip between the knife and the toast," she says. "Our honey is very, very thick. We leave it on the hive until the moisture content is really low." The lower the moisture the better the quality. Canada's best is 17 percent. Babe doesn't put out honey that's over 15 percent. Part of its attraction is appearance. Fireweed produces honey that seems as golden as sunlight. Wild blossoms on salmonberries, salal, oregon grape and alpine meadow flowers yields honey of a tawnier colour with deep amber glints of an intrinsic beauty.

Babe won't say what her production has been this year—"that's strategic information"—except to say the wet weather is the worst in fifty-two years and the flow was down by 70 percent. "In 1997 I had a ten-month market," she says. "This year I doubt we'll get six months. When we run out, that's that. I don't buy and pack under my label. Only my honey goes out of here." That means a sales dilemma. She could unload everything at a higher retail price over the farm gate to customers who drive in from as far away as the States. "But we've been supplying local grocery stores for 48 years. What about our loyal customers who are too old to come out here. We have to supply them, too."

That, of course, is precisely why she has such loyal customers, a kind of metaphor for the enduring relationships between city and country, between the people and the land, to which our Thanksgiving celebration pays homage. These relationships between growers and consumers, processors and distributors, shippers and retailers, all add up to a BC food business that now tops $16 billion a year in value. Not bad for a place where "Fifty railroads could not galvanize it to prosperity," as the British newspaper *The Truth* predicted just over one hundred years ago.

Land of Dreams and Miracles

Almost five generations ago, the Nisga'a chief Na-qua-oon was travelling down the turbulent Nass River that is the spiritual conduit between his people and their past. Once rich beyond belief in salmon and game, the river sweeps down from glittering ice-clad mountains. They crumple skyward out of the collision between ancient volcanic seamounts and the primordial rock of the North American plate. Eight million years older than the Himalayas, these peaks still ascend like an express elevator, in some places rising by more than the tallest man's height every century. The Nass is contemptuous of time. It carves an urgent path through the geological past, laying bare in its canyons the record of cataclysm and upheaval. Farther down, it wanders peacefully toward its estuary through alluvial flood plains shaded by the rustling canopies of cottonwoods and other trees that were majestic when Captain Cook was a lad.

In geological terms—spans which exceed even the time-shrouded oral histories of the Nisga'a—this is Wrangellia, part of a buckling, shuddering subduction zone of exotic, displaced rocks that reaches south as far as Vancouver Island. Rich in minerals, it is a dramatic world of elementals, of great earthquakes and tsunamis, of flood and fire, of rivers of magma welling up through ruptures in the earth's crust. At the entrance to the Nass Valley, the traveller encounters the stark vista of Anhluut'ukwsim Laxmihl Angwinga'asanskwhl Nisga'a, the jumbled lava bed where a memorial park commemorates 2,000 people entombed one molten night in 1,300-degree rock. That eruption is recorded in the history of the newcomers, too. We can date it to 1775, when Padre Miguel de la Campa noted in his diary that the crew of the Spanish ship *Sonora* suffered from the heat from "the great flames which issued from four or five mouths of a volcano and at night lit up the whole district, rendering everything visible." To experience this country, to smell its pungency, to feel its bite, to engage it fully with the senses rather than the intellect is to comprehend at once how the Nisga'a are irrevocably bound to their homeland.

Na-qua-oon, whose name means Long Arms—he whose generosity is great enough to embrace the whole tribe—was born in 1869, just about the mid-point of European history on this coast. The newly minted colony of British Columbia had not yet joined confederation. Canada itself was an infant. His own elders could still remember the first coming of the K'umsiiwa, the white people. The word means "driftwood"—an apt term that points both to the subtle nuance of the Nisga'a tongue and to the dry humour of its speakers. "K'umsiiwa" implies a duality of character, referring not only to the pale colouration, but also to the rootlessness of the newcomers, drifting thither and yon on the busy tides of their acquisitiveness. Indeed, the debris of abandoned canneries, the shambles of decayed mining camps like nearby Anyox, even the seedy, ramshackle transience of the present-day logger's community at Nass Camp, all substantiate this sardonic assessment.

Na-qua-oon was hereditary chief of the Wolf Clan, one of the four most ancient clans of the Nisga'a, a people who have occupied the Nass since the ice was two kilometres thick and most of North

America was as silent and devoid of life as Antarctica. And if the mythic origins of his clan are entangled with the wolf, it is worth recalling that among the most formidable ice age presences was the now-extinct dire wolf—a creature to make today's timber wolf look like a house pet. Yet during his journey down the Nass so long ago, all the power of Chief Na-qua-oon proved helpless in preventing his hopes for the future from being extinguished. His young son was drowned in the river. It must have seemed a crushing portent. Since the coming of the K'umsiiwa, epidemic after epidemic had reduced the sixteen Nisga'a villages to four. From a population of perhaps ten thousand, the clans had dwindled to 738 survivors. Denied their historic trading rights by government order, the very people who felt the land part of their collective soul had been turned into industrial nomads. Some went logging. Some followed the cannery trail. Chief Na-qua-oon himself left the Nass to seek work, labouring to clear a raw new townsite in the south and pitching his tent where Vancouver's Georgia Hotel now stands.

Then, with mysterious force, came the dream. First to an old woman at Kincolith, the Nisga'a village at the mouth of the river, a woman so old even then that those who might remember her name are now mostly dead. Her dream was sufficiently vivid that she went out and stopped a young man on the beach. She insisted that he row her the 10 kilometres to Chief Na-qua-oon's house. There she spoke with the chief's wife, Louisa. Over in Nass Harbour, she said, Louisa's youngest sister, Emily Clark, would soon conceive. The baby would be a boy. Into that yet unborn child would enter the chiefly spirit of Na-qua-oon's drowned son. And so it was that in the summer of 1915, when Emily presented her husband Job with the son of whom the old woman had dreamed, Chief Na-qua-oon and Louisa adopted him according to Nisga'a law.

Four years later at Kincolith, all the chiefs had gathered to address the problem that was proving an immovable mountain: the land question—the K'umsiiwa's desire to disconnect the Nisga'a from their land and the chiefs' refusal to yield. At this feast, Chief Na-qua-oon, who spoke little English, picked up his son and with his powerful arms held him out before his fellow chiefs. He told them then of his own dream: "This boy is going to learn the language and the laws of the K'umsiiwa. When he comes back, he's going to move that mountain." Na-qua-oon's English name was Arthur Calder. And thus, from a landscape of dreams and prophecies, began the long and arduous journey of Frank Calder.

That child of the old woman's dream became one of the first aboriginal university graduates in Canadian history. He became the first aboriginal person in the Commonwealth elected to a legislature, serving twenty-six years in public office. He was the first aboriginal person appointed a minister of the crown. Perhaps most important, he helped found the Nisga'a Tribal Council and, true to his father's promise, resurrected the vexing land question. It was his lawsuit against the crown that finally forced obstinate governments to engage at last in negotiating a just treaty with the Nisga'a.

That was in 1969. War raged in Vietnam. Student radicals stormed the campuses. Violence rocked Quebec. The Soviets were crushing the Prague Spring. Human beings walked on the moon. Nobody knew or cared about a small people's quiet fight for justice on the northwest shoulder of British Columbia. The Nisga'a had been demanding recognition of their title to lands they occupied for thousands of years before European settlers arrived with new laws and institutions. Petitions and pleas and lobbying had fallen on barren ground for more than a century.

Frank Calder was a feisty divinity student turned politician when he made the Nisga'a a lightning rod for the aboriginal cause across Canada. Represented by lawyer Thomas Berger, Calder was the young president of the Nisga'a Tribal Council when he sued the provincial government and asked the court to affirm unextinguished Nisga'a title to traditional territories. The Nisga'a argument leaned heavily on the Royal Proclamation of 1763 in which the

British Crown acknowledged aboriginal title and insisted colonial governments make treaties to extinguish it. BC countered that the Royal Proclamation could not logically apply to the then undiscovered regions of what is now the province.

Like so many aboriginal claims, the suit was dismissed in provincial court. It lost again when the BC Appeal Court ruled that the Nisga'a had legal title to traditional lands—but only where that title was explicitly recognized by some act of government, a recognition which had never been acknowledged by the province. Isolated and repudiated, the Nisga'a refused to quit. Calder and his council instructed Berger to appeal to the Supreme Court of Canada. In 1971 the highest court agreed to hear the case. In 1973 it brought down a verdict that rocked BC's constitutional assumptions. Six of the seven judges ruled that the Nisga'a had held aboriginal title to traditional lands. But the court split evenly on the question of whether aboriginal title continued after colonial governments were established. Three judges said it did. The seventh judge ruled that the Nisga'a had no technical right to sue the crown and never dealt with the issue of aboriginal rights.

Technically, the Nisga'a had lost their case. Morally, they had won a great victory. Not only that, in his minority opinion, Justice Emmett Hall dismissed arguments that the intent and force of the Royal Proclamation of 1763 did not apply to BC with such a brilliant and withering historical analysis that it is still frequently cited. Sir Francis Drake had laid claim to the west coast in 1579, he pointed out. The British knew the Russians were trading on the northwest coast by 1742. De Niverville's expedition to the Blackfoot in 1751 reported that Indians to the west were trading with white men on the far side of the Rocky Mountains. Hall observed that the Proclamation itself said that it applied to "all the lands and territories lying to the westward of the rivers which fall into the sea from the west and the northwest." How in any rational argument could this be seen to exclude BC? The government of

Canada responded immediately with a comprehensive land claims policy to renew the treaty-making process and on January 12, 1976 began negotiations with the Nisga'a. An obstinate BC government, still mired in the colonial assumptions that had permitted the plundering of natural resources and land grab after land grab from aboriginal occupants, stalled for another fifteen years before joining the negotiations. It was to be another five years after that before an agreement in principle was achieved.

In the meantime, Frank Calder and the rest of his generation of Nisga'a leaders have become old men. Many have died. But this small and determined nation, passing the torch from father to son, mother to daughter, had indisputably changed the course of Canadian history. And yet, vocal opponents of aboriginal self-government, heirs of the colonial attitudes of nineteenth-century imperialism, finding political allies among the resource interests and developers, still wage a tireless campaign in the media to try to kill the fruits of that process. Despite that noisy opposition, after an overwhelming majority of the 6,000 Nisga'a endorsed an agreement in principle, federal and provincial governments negotiated and signed a treaty and in 1999 began procedures for having it formally ratified by their elected legislatures.

To some it seems a miracle, the realization of a dream which began more than a century ago when Nisga'a chiefs politely rebuffed government surveys and insisted the Nass was theirs. "We have survived," a young artist named Dennis Nyce told me as we squinted into a cutting wind off distant snowfields. "The journey our forefathers started—we've survived to see that canoe come back after its long journey to the land of the K'umsiiwa." Frank Calder, who remains partly rooted in the land of the K'umsiiwa, who was placed in that canoe as a toddler by a father who did not live to watch it set out, let alone return, sees the beginning of his people's journey, not the end of his own. "Now we're going to move another mountain," he says. "And this time all the Nisga'a are going to move it together."

The mountain this time means finding a way to merge modern democratic institutions with the Ayuukhl, the ancient, unwritten body of laws which regulate all Nisga'a social activity. Frank Calder is well past eighty and his accumulated wisdom seems unassailable. In some ways, he says, the hardest journey is the one yet to come, the work of forging a self-sustaining homeland from the crumbs left by the K'umsiiwa. How does a society move from the politics of kinship to the politics of electoral plurality? What structure should replace the imposed oppression of the Indian Act? How do hereditary chiefs preserve what it is to be Nisga'a when the forces of globalization drive us into the worldwide village? In a way, the landing of the canoe launched by Na-qua-oon marks only the end of the beginning. A new canoe must carry Nisga'a aspirations through the turbulent and uncharted waters of discovery that the treaty represents.

Yet there are glimpses of what might be. In Nisga'a country, when you get the grand tour from proud civic officials, you are not shown the impressive new hockey rink or the wonderful mall or the awe-inspiring city hall. Here the public architecture speaks to another set of priorities. At New Aiyansh, the seat of an emerging Nisga'a government is housed in a battered collection of trailers with uneven floors. Between meetings, its unpretentious boardroom is littered with working maps from the resource database that will form the council's geographic information system and the executive director occupies an office that a high school custodian might complain was too small.

In New Aiyansh, as in Gitwinksihlkw and Lakalzap, the public money goes not to the comfort of politicians, but to the building of new schools and health units and homes for the citizens. The best view in New Aiyansh goes to the patients and staff at the shiny new health centre. At Gitwinksihlkw, the new school dominates the civic landscape. It resembles a longhouse. At the centre, in place of the traditional hearth, is the library and computer lab. Outside is one of the two new totem poles in the community, a statement both of cultural renewal and of the school's symbolic importance to the people who built it. Inside, a more literary statement:

> In the beginning was a dream. And the dream was in the hearts of our people. And this was our dream: to have our very own school in the bosom of our community which would serve as a Treasure House where-in to entrust the Sacred Wisdom of our elders.
>
> This beautiful edifice will not guarantee us education. True education resides in our hearts and our minds. We educate best when we raise our children in the love of Beauty: with Beauty inside us and Beauty outside us; with Beauty above us and Beauty below us; with Beauty all around us, travelling towards our Destiny on the Beautiful Road of Life!

In New Aiyansh, in a small computer laboratory attached to the tribal council's offices, Gary Patsey and a team that looks barely out of its teens is busy analyzing downloaded satellite images. From these they fashion a highly detailed inventory of the timber resources inside the proposed Nisga'a homeland. These maps will be part of the first overt act of decolonization—the restoration of traditional Nisga'a place names. "I know the name of every place in this valley. I know the name of every mountain, every hill, every stream and every spring. Restoring those names is one of our priorities," says Rod Robinson, a chief of the Eagle Clan and executive director of the Nisga'a Tribal Council. Digital satellite images and software engineers, boundaries and names that originated during the last ice age—the mind lurches. This is indeed a land of bewildering contrasts. It is a place where dream and reality merge into a seamless equilibrium of past and present. It is also a place where striking beauty is tainted by unpardonable avarice.

To get to the country of the Nisga'a, you take a plane to Terrace, a raw-boned, newly uncrated resource town 1,355 kilometres northwest of Vancouver. Then you drive north, up, over the gentle pass between Sleeping Beauty and Lean-to Mountain and then 100 kilometres out into the bush. The pavement soon gives way to soupy gravel and every blind corner threatens a loaded logging truck. On the way in, I counted a truck every seven minutes. Their cargoes were prime old-growth trees. Leaving the Nass to drive back, empties headed into the valley at roughly the same rate. While the Nisga'a negotiated their treaty, this conveyor belt ran twenty-four hours a day. And the team creating the computerized resource inventory says the rate of cut accelerated as a negotiated settlement of the land question drew near. How, one is compelled to wonder, do we explain the torrent of wealth that steadily poured out of traditional Nisga'a territory even as governments negotiated a treaty validating the tribal council's continuous legal ownership of those resources? "High-grading," Frank Calder calls it. Treaty or no treaty, throughout the valley, the shoulders of the mountains are still scabbed by vast clearcuts and the salmon runs are in decline. Logging companies clamour for cash compensation. Treaty foes rail about aboriginal fishing provisions.

Fourteen years ago, the tribal council commissioned a study of timber resources in the Nass Valley. Up to 96 percent of the area logged and said to be restocked was found not satisfactorily reforested. It found that 73 percent of the best quality logged areas were choked with brush, that 25 percent of the soil had been severely degraded and that the province had deliberately relaxed its normal contractual requirements for reforestation, timber utilization, cutblock planning and environmental protection. In 1985, a special report by the provincial ombudsman found that the BC government had acted improperly in failing to enforce forestry standards, had failed to meet its statutory obligations to manage and conserve the resource and had simply broken the law when it saw fit to do so. "The way the land is at this point in time—it lies devastated," says Robinson. "Who will pay to repair it? What appals our people is the demand by Repap for $80 million in compensation. What about the damage that has been done that requires four times that amount to reverse it?" It is difficult to fault his logic. Coming into the country of the Nisga'a may be to experience the intersection of dreams and miracles, but it is also to witness an astonishing symbiosis of greed and cynicism. Unleashed by the dominant society, these cruel twins devour the resources of a ravaged landscape like mythological monsters out of aboriginal myth. "They have no religion, they have no creed, they have no race—their only value is greed," says Frank Calder. "These are the people who demanded a place at the table. I kept telling our people: 'No, you don't allow those guys in there.'"

Small surprise, then, that people seem uncertain whether to exult or despair over an agreement which now seems destined to set a template for future treaties between Canada, British Columbia and the province's First Nations. "It is a bittersweet fruit," says Maurice Squires, director of the Nisga'a family and child care program. "There are people who are in grief. We gave up so much." Some villages feel excluded by the whole process. The traditional lands of one hereditary chief, for example, lie outside the heartlands covered by the treaty framework. These lands have been his responsibility since time immemorial. What does their exclusion mean—that he is now a lesser chief? Collectively, the Nisga'a hope not. They hope that negotiating a treaty means they are once again in control of their history. They believe they are negotiating a way back into the Canadian body politic that will protect and defend a collectivity that for the past one hundred years we in the dominant culture have done our best to destroy.

Yet these dreams seem more than the present political hopes of a small people seeking self-determination in the shadow of a homogenizing leviathan. They represent something larger, a hidden spiritual pulse at the core of a profound culture which struggles to greet the future while keeping faith with its origins. Talk to elders or to tribal politicians or to young people you encounter by chance on the dusty streets—references to dreams enter almost every conversation. At times, they seem a mysterious portal through which a mythic past discharges meaning into our exacting world of reason and the narrow rationalizations of scientific materialism. The greatest gift to us from the Nisga'a is the chance to see in their experience something of our own repressed origins. In human terms, the Nass River valley is one of the most ancient places in what is now Canada, continuously occupied by the Nisga'a for at least 8,000 years, probably 10,000, perhaps twice that. The same people practised the same law and the same forms of government, used the same words to tell the same stories, harvested and prepared the same natural bounty while elsewhere Ozymandias and Babylon rose and fell and were forgotten.

How long have the Nisga'a been there? Where do they come from? This seemingly innocent question arises frequently in the discussion over land claims. But Maurice Squires bristles at the question. To him it is a transparent implication that the Nisga'a are simply immigrants with no greater claim than the rest of us. "That whole concept of us coming across the land bridge from Asia is just another way of manipulating individuals to believe that I am not from here. But I am." His logic seems reasonable. Outside the framework of anthropological theory, the question seems foolish, like asking "How long have the Chinese been there? Where did they come from?" Or "When did the British cross the land bridge from Europe?" The answer is that just as we have no idea who held power in Britain 10,000 years ago, nobody knows how long the Nisga'a have lived on the Nass or where they came from—which validates their own view that they have been there forever.

Nineteenth-century anthropologists who had never visited the Nass classified the Nisga'a as a cultural subgroup of the Tsimshian, an assumption which the Nisga'a themselves reject, sometimes with vehemence. Outside the tribal supermarket, a fisheries poster warns about contaminated shellfish from Digby Island:

"Don't eat these clams!" Across it some wag has scrawled "Eat the Tsimshians!" It's true that a third of the words in Nisga'a language appear to derive from the Tsimshian language stock, but the present day inhabitants argue that they were a great trading people—to trade, you absorb the language of your main client.

In the crowded parlour of tribal elder Emma Nyce, a devout woman, mother of twelve children, sister of the late chief James Gosnell, the language is that of parable and legend. Her house is built on the same site as the longhouses of Gisk'ahast, Gitwilnak'il, Ganada. Inside the air is heavy with tradition. On the walls, a gallery of photographs complements the genealogies in her memory. And she can remember when there were only four houses where a village of two hundred people now thrives. Her grandmother was ninety-six when she died in 1948.

"The old women—they told us the law of the Nisga'a," she says. "How we were to be married. How we were to be buried. The law of the people.

"My grandmother said to me, 'You're not going away. You are going to get married. You are going to live here and have children here.' I was married by Nisga'a law. I've lived here all my life. My children were born here without hospitals. My mother and father delivered my children. This is our family's place." She, too, is steeped in dreams and stories of dreams.

Across the green coils of the Nass, across the jagged lava beds, beyond the cave where girls at puberty would spend a year preparing themselves for womanhood, rises the great mountain K'lawit, to which is tethered the whole history of the Nisga'a. It was here that the people found sanctuary, mooring their canoes with cedar ropes to ride out the world-swallowing flood. A dim memory of some great glacial melt as the ice age retreated? Or a parable for the tenacity of a people's hold on a cherished place? Or simply the truth, as true as any other remembrance of Eden? I asked Frank Calder his opinion.

"My father told me," he said. "The people could feel that there was going to be a flood. They couldn't save their villages but they knew that their land couldn't be swept away. They stayed. They didn't want to lose sight of their valley.

"When the water receded, they went back and established the villages that are there today. That's who we are. We live on dreams."